OFDM for Wireless Communications Systems

For a listing of recent titles in the *Artech House
Universal Personal Communications Series,* turn to the back of this book.

OFDM for Wireless Communications Systems

Ramjee Prasad

Artech House, Inc.
Boston • London
www.artechhouse.com

Library of Congress Cataloging-in-Publication Data
Prasad, Ramjee.
OFDM for wireless communications systems / Ramjee Prasad.
    p. cm—(Artech House universal personal communications series)
Includes bibliographical references and index.
ISBN 1-58053-796-0 (alk. paper)
1. Wireless communication systems.   2. Multiplexing.   3. Orthogonalization methods.
I. Title.   II. Series.

TK5103.2.P715 2004
621.382—dc22   2004053828

British Library Cataloguing in Publication Data
Prasad, Ramjee.
OFDM for wireless communications systems—(Artech House Universal Personal
Communications series)
1. Wireless communication systems   2. Multiplexing
I. Title
621.3'84

ISBN 1-58053-796-0

Cover design by Yekaterina Ratner

International Standard Book Number: 1-58053-796-0

10 9 8 7 6 5 4 3 2 1

To my and my wife Jyoti's lovely granddaughters Sneha and Ruchika, whose innocent, smiling faces keep us energetic

Contents

CHAPTER 8

Preface

सहजं कर्म कौन्तेय सदोषमपि न त्यजेत् ।
सर्वारम्भा हि दोषेण धूमेनाग्निरिवावृताः ॥

saha-jaṁ karma kaunteya
sa-doṣam api na tyajet
sarvārambhā hi doṣeṇa
dhūmenāgnir ivāvṛtāḥ

Every endeavor is covered by some fault, just as fire is covered by smoke. Therefore one should not give up the work born of his nature, even if such work is full of fault.
—*The Bhagvad-Gita* (18.48)

My wireless (mobile) garden is full of flowers with varieties of flavors, for example, CDMA, OFDM, and so forth.

Last year I realized my "OFDM flower" has become a "paragon" with the successful completion of the work of of several masters, doctoral, and postdoctoral candidates, for example Mohindar Jankiraman, Dusan Matic, Klaus Witrisal, Uma Jha, Richard van Nee, Shinsuke Hara, Hiroshi Harada, and so on. Therefore, I decided to put together in one place their interesting and valuable contributions, particularly of those of Klaus, Jankiraman, Uma, and Richard.

Although I coauthored *OFDM for Wireless Multimedia Communications* with Richard and *Multicarrier Techniques for 4G Mobile Communications* with Shinsuke, this book is very different in that it presents an overview of the wireless local area network (WLAN), wireless personal area network (WPAN), frequency-domain channel model, a novel hybrid OFDM concept, and a practical OFDM system.

Figure P.1 illustrates the coverage of this book. This book illustrates the role of OFDM in developing an adaptive system by designing OFDM-based wireless wide area networks (WWANs), WLANs, and WPANs. It is based on the contributions of several researchers who had or have been actively involved in growing the OFDM flower in the wireless (mobile) garden under my gardenership.

As a gardener, I have tried my best to provide enough water and energy to nurture the OFDM flower up until this point. In the future, it will sow several other interesting colors, which I will bring to you at that time.

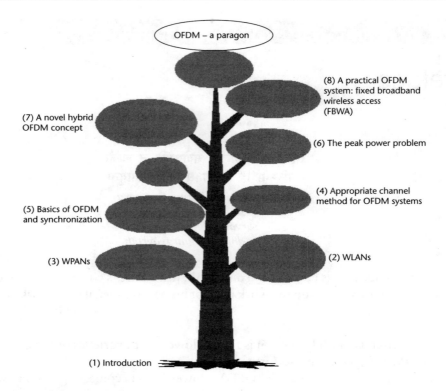

Figure P.1   Coverage of this book.

I would greatly appreciate it if readers would provide extra water and energy in improving the quality by pointing out any errors. I strongly believe nothing is errorless.

Acknowledgments

I would like to express my heartfelt gratitude to colleagues and students without whom this book would have never been completed, namely, Mohindar, Dusan, Klaus, Uma, Richard, Petar, Hiro, Carl, Liljana, Shinsuke, Hiroshi, and Anand.

Junko gave her support in preparing the typescript of the book.

CHAPTER 1

Introduction

"It is dangerous to put limits on wireless data rates, considering economic constraints," I said in 1999. Data rates are really what broadband is about. Broadband wireless communications will support applications up to 1 Gbps and will probably operate in the 60-GHz frequency [1–5]. However, many people argue whether there is a need for such high-capacity systems, bearing in mind all of the compression algorithms developed and the types of applications that require tens of megabits per second. One can look at this issue from another perspective. There is a need for high-capacity systems to give a perspective of what should be the "hot topics" in the area of telecommunications for research. In this visionary perspective of the road to follow, in order to go along with the needs of society in the year to come as far as communications is concerned, capacity is one of the major issues to be developed due to the foreseen increase in demand for new services (especially those based on multimedia). Together with this, personal mobility will impose new challenges to the development of new personal and mobile communications systems.

A conclusion can be drawn from this: Even if at a certain point it may look "academic" to develop a system for a capacity much higher than what seems reasonable (in the sense that there are no applications requiring such high capacity), it is worthwhile to do it since almost certainly in the future (which may not be very far off) applications will come out that need a capacity of even more than 1 Gbps. The story of fiber optics is elucidative on that. Rapid development will shrink the world into a global information multimedia communication village (GIMCV) by 2020. Figure 1.1 illustrates the basic concept of a GIMCV, which consists of version components of different scales ranging from global to picocelluar in size. Figure 1.2 shows a family tree of the GIMCV system [6–21].

1.1 Wireless Technology in the Future

Today, basically five wireless technologies have made an impact, namely, wireless global area networks (WGANs), wireless wide area networks (WWANs), wireless local area networks (WLANs), wireless personal area networks (WPANs), and wireless broadband–personal area networks (WB-PANs), as illustrated in Figure 1.3.

These five technologies will not compete with, but will complement, each other.

Another set of technologies is fixed wireless access (FWA) or broadband wireless access (BWA). Current standardization trends show that the FWA technologies will get mobility functionalities; if this happens, then FWA could become a threat to

1

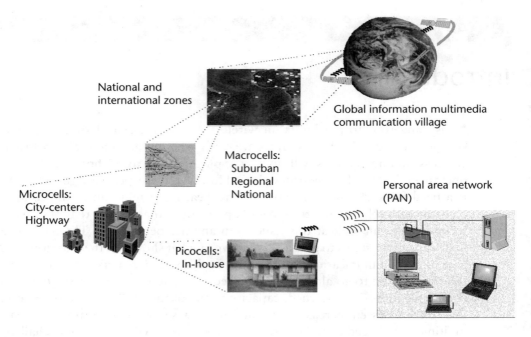

Figure 1.1   GIMCV.

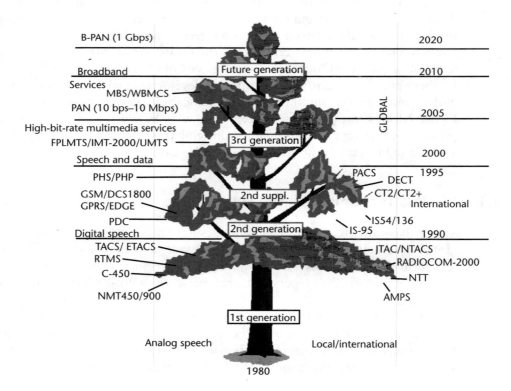

Figure 1.2   Family tree of the GIMCV. Branches and leaves of the GIMCV family tree are not shown in chronological order.

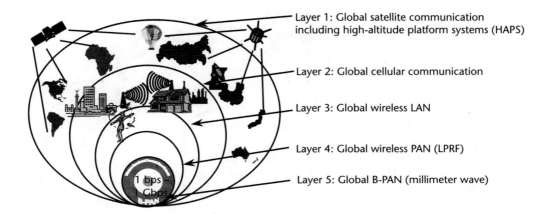

Layer 1: Global satellite communication
including high-altitude platform systems (HAPS)

Layer 2: Global cellular communication

Layer 3: Global wireless LAN

Layer 4: Global wireless PAN (LPRF)

Layer 5: Global B-PAN (millimeter wave)

Figure 1.3   Five-layer wireless communications provide mobile everywhere and complement each other.

WWANs. Development of 802.20, a mobile BWA (MBWA), could surely become a threat for WWANs in the future. In the following the future direction of WLANs, WWANs and WPANs is presented; Table 1.1 presents an overview of wireless technology standards. Figure 1.4 shows the partitioning among WWAN, WLAN, and WPAN.

1.1.1   WWANs

Growth in the WWAN field, more commonly known as mobile communications, has been tremendous over the past decade. Second generation (2G), 2.5G, and third generation (3G) standards of mobile systems are being used, while efforts are ongoing toward development and standardization of beyond 3G (B3G) systems. The existing 2G systems are mainly for voice purposes. Due to the tremendous growth of the Internet, some support for data services like Wireless Application Protocol (WAP) and I-mode have been developed [22, 23]. Further, 2G supplement systems, 2.5G,

Table 1.1   Wireless Technologies

Cellular Technology	WLAN	WPAN	Cordless Technology	FWA/BWA
GSM-HCSD, GPRS, EDGE-(WAP)	IEEE 802.11	IEEE 802.15	PHS	IEEE 802.16, IEEE 802.20 (MBWA)
IS-95	HIPERLAN/2	Bluetooth	DECT	HIPERACCESS
IS-54/IS-136	MMAC Ethernet WG and ATM WG (HiS-WAN)	HIPERPAN	CT2/CT2+	High-speed wireless access
PDC (I-mode)	MBS			BWIF
3G	MMAC wireless homelink			LMDS
				MMDS

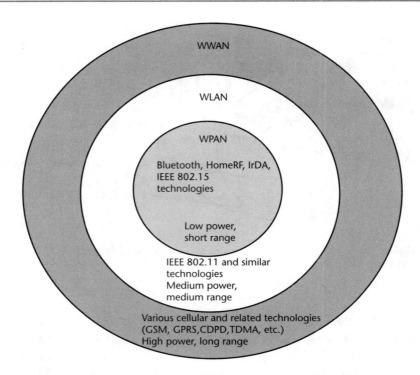

Figure 1.4 Partitioning among WWAN, WLAN, and WPAN.

like General Packet Radio Systems (GPRS), and now 3G systems provide further possibilities for data services with varying quality-of-service (QoS) requirements.

At present the main application for data services over mobile communications systems is Internet access. The future is toward full multimedia-type applications providing various levels of QoS using an Internet Protocol (IP)–based backbone. Thus, WWAN is also moving toward integration of services.

Further works are being done by the standardization committees to integrate WLANs with 3G. Another development is the standardization of WWAN toward an IP network. All this shows us that the WWANs are moving toward packet-switched solutions and the integration of technologies, now that the integration of services is almost achieved.

1.1.2 WLANs

Local area networks (LANs) mostly make use of IPs. The growth in wireless and the benefits it provides have brought forward changes in the world of LANs in recent years. WLANs provide much higher data rates as compared to WWANs for slow mobile or static systems. Institute of Electrical and Electronics Engineers (IEEE) 802.11b–based WLANs are already widely being used while IEEE 802.11g and IEEE 802.11a are also available on the market.

WLAN technologies are mainly used for wireless transmission of IP packets. Until now, in contrast to WWANs, WLANs have provided network access as a complement to the wireline LANs. In the near future QoS-based WLANs are expected to come onto the market.

IEEE 802.11e is working toward medium access control (MAC) enhancements. The purpose of the MAC enhancement is to enable the present MAC, CSMA/CA, to provide QoS. The current draft has accepted two variations for QoS enhancements: These are central-control- and distributed-control-based. For security, in IEEE 802.11i, the main direction is toward applying IEEE 802.1X-like solutions with stronger, and more varied choice of, encryption algorithms. The IEEE 802.11 working group (WG) has also accepted a mobility solution known as Inter Access Point Protocol (IAPP), IEEE 802.11f. Another group in IEEE 802.11 is working on radio resource management (IEEE 802.11j). The IEEE 802.11 committee has approved IEEE 802.11h, dynamic frequency assignment and transmit power control. Due to the success of the standard, several other study groups are looking at higher-data-rate solutions (IEEE 802.11n 110 Mbps+) and next generation technologies, including standardization work with 3G standardization committees.

The WiFi Alliance, an industry alliance, is providing interoperability specifications and tests of the IEEE 802.11 products for better acceptance by the market. This alliance also provides recommendations for roaming between different wireless Internet service providers (WISPs) so that the customer of one WISP can access WLAN services when in another WISP's hotspot and still receive one bill.

Other known WLAN technologies are HIPERLAN Type 2 and HomeRF. HIPERLAN Type 2 is already standardized; it provides hooks for QoS and security for different environments. HomeRF developed several solutions, but in early 2003 was proclaimed dead.

The direction for WLANs at present would be to move toward a common international standard. Harmonization of the 5-GHz band technologies is a must so as to avoid making the 5-GHz band a garbage band. Although harmonization is a solution, it is possible that the market will be a deciding factor and choose one technology. For the time being the success of a standard will depend on the pricing, performance [24], availability, and marketing of the standards.

Besides the work being done by the standardization committees, studies should be made on providing top-to-bottom mapping. The correct mapping of higher-layer protocols to lower-layers protocols is a must to provide optimum service. Especially in the case of IEEE 802.11, where the standard only defines the bottom two layers, relations must be created with Internet Engineering Task Force (IETF), the committee developing layer-3 and some higher-layer protocols.

Basically, most of the current developments will lead to providing users with different services using WLANs, or in other words, toward integration of services within one system. Another step currently becoming visible is toward integration with WWAN technologies like 3G.

1.1.3   WPANs

Besides the WLANs, the WPANs like Bluetooth, HIPERPAN, and IEEE 802.15 have been standardized. These technologies will be used for short distance (~10m) communications with low data rates for different QoS. It is envisaged that the

WPANs will exist in all of the mobile terminals in the near future. The WPAN standards, IEEE 802.15.3 and IEEE 802.15.3a, have developed and work is ongoing to develop higher data rates of about 55 Mbps, paving the path toward broadband WPANs. IEEE 802.15.4 is focusing on very low data-rate solutions, which will work at a few or a few hundred kilobytes per second; this is a first step toward the development of body area networks. Several companies have reached consensus on ultra wideband (UWB) as a low-data-rate solution for IEEE 802.15.

1.1.4 WB-PANs

The WB-PAN is a future development of PAN toward the wideband-adaptive novel techniques capable of broadband wireless communication. It will support applications of up to 1 Gbps and will probably operate over the 5-GHz or 60-GHz frequency bands [1]. WBPANs will implement novel techniques such as UWB, voice over WB-PANs, smart antenna, adaptive modulation, coding, and the like, with extendable protocol functionalities. It should support performance QoS in an adaptive and flexible manner. Different access methods and application interfaces will be defined, and the system will be supported with segmented intelligent multiaccess terminals capable of speech, messaging, and multimedia operations.

The WB-PAN belongs to the wireless family, appearing to be one of the most promising concepts, which opens tremendous possibilities for new applications. Table 1.2 presents the technical differences between several previously mentioned wireless systems.

Table 1.2 Technical Differences and Applications

	UTRA	WLAN	Bluetooth	PAN	WB-PANs
Data rates	Maximum 2 Mbps (384 Kbps)	5.1–54 Mbps	Maximum 721 Kbps	Max. 10 Mbps	1 Gbps
Technology	TD-CDMA and WCDMA	OFDM	DS or FH	OFDM	OFDM/DS-CDMA/SHF-CDMA
Cell radius	30m–20 km	50–300m	0.1–10m	To the distance an voice reaches	Similar to PAN
Mobility	High	Low	Very low	Very low	Very low
Standard availability	1999	2000	1999	2004	2012
Frequency band	2 GHz	5 GHz	2.4-GHz ISM band	5–10 GHz	60 GHz
Frequency license	Necessary	Not necessary	Not necessary	Not necessary	Not necessary
Application	Public environments (likely restricted use in hospitals, on airplanes)	Corporate environments (industrial applications); public hotspots (airports, exhibitions, convention centers)	Substitution for infrared communications; low cost networks for SoHo and residential applications	Personal peripheral device communications	Surrounding environment

1.1.5   The Next Generation

Each wireless technology is moving toward future standardization. This standardization work is mainly focusing on wireless IP-based QoS provision for any type of data. Here data comprises everything, be it audio, video, gaming, or any other application. Basically, this means an integration of services. All these technologies' areas of service overlap to some extent. This is illustrated in Figure 1.5. Thus, a move toward integration of technology is a logical next step to provide service continuity and higher user experience (quality of experience) (see Figure 1.6).

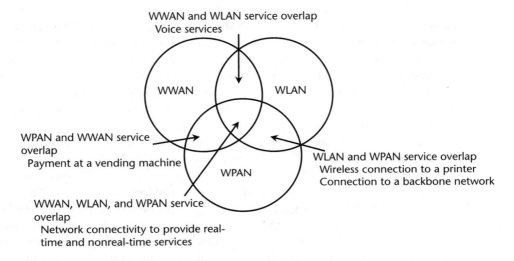

Figure 1.5   WWAN, WPAN, and WLAN overlap.

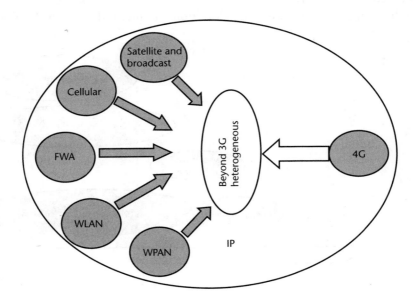

Figure 1.6   Future of telecommunications.

The ITU-R vision for 4G also calls for integration of technologies, commonly known as heterogeneous systems or B3G systems (to some people B3G could mean any standard or technology developed after 3G). Integration of technology will provide adequate services to a user depending on mobility and availability. Of course, this brings along several new challenges, for example, handover/handoff or mobility, security and QoS. These issues should be resolved without changing the existing standards. Seamless handover should be provided while a user moves from the network of one access technology to another and the domain of one stakeholder to another. Seamless handover means provision of seamless service while the user is mobile; that is, the user does not perceive any disruption in service or quality during handover. The IEEE 802 Handoff Executive Committee Study Group (ECSG) is working on the issue of handover for 802-based technologies.

The ITU-R vision also talks about a new air interface, also known as 4G. As any new system takes about 10 years to develop and deploy (see Figure 1.7), work on B3G and 4G has already started, and a possible solution is given in [25]. The current market shows that 3G is having trouble, and hopefully the lessons learned will be taken into account in the development of 4G [26].

A possible future scenario is given in Figure 1.8. All technologies should work together while providing all of the services to users anywhere anytime. Table 1.3 shows the envisaged development in the stakeholders of the various networks and technological development for the short-, mid-, and long-term future. The table also points out several technological issues that should be worked on. Arrows between two cells of the table represent the possibility of handover between the two technologies, while the shade of the arrow (shown in grayscale) represents the expected extent of the handover. Research work should be done on seamless handover, which brings in the study of several issues like security and QoS, which should be conducted at each protocol layer and for each network element. This topic itself will require further study of development methods and technologies, including

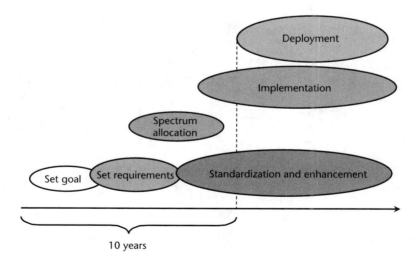

Figure 1.7 Time required for new technology development and deployment.

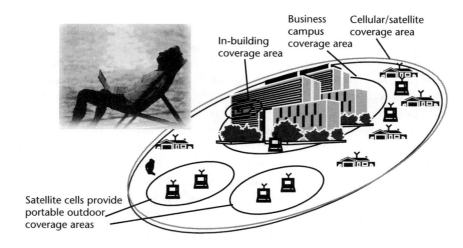

Figure 1.8   Future of wireless.

Table 1.3   Envisaged Technology Development in the Short, Mid, and Long Term

NG (+) Next generation and beyond handover

	Stakeholder (of one or more access networks)	IP	Broadcast (DVB-T, DAB)	WWAN (3G & 2.5 G)	WLAN (IEEE 802.11)	FWA (FWA, ISDN)	WPAN and Adhoc (IEEE 802.15)
Short term (2–3 years)	Same (not broadcast)	v4	Similar to TV or radio	3G & 2.5 G handover: maybe	b, g, a		
Mid Term (3–5 years)	Same (maybe few different, surely not broadcast)	v4&6	As above	3G → 3.5G and 2.5, handover: possible	g, a, NG, QoS		
Long Term (5–10 years)	Same and different	v4&6	SDR	2.5G, 3G, 3.5G and 4G handover: must	g, a, NG, QoS		

Gray   For handover: Three levels of gray signify the expected extent of handover. The darker the arrow, the more common the handover between the concerned technologies.
    For technologies (e.g., 2.5G): Shades of gray signify decreased use of standard.

hardware, software, firmware, and technologies like application specific integrated circuits (ASICs). Another important research topic is software-defined radio (SDR), which includes reconfigurability at every protocol layer.

WLANs provide roaming within LANs, and work is ongoing toward further enhancement in this field. While WWANs provide roaming, too, the challenge now is to provide seamless roaming from one system to another, from one location to another, and from one network provider to another. In terms of security, again, both WLANs and WWANs have their own approach. The challenge is to provide the level of security required by the user while he or she is roaming from one system to other. The user must get end-to-end security regardless of the system, service provider, or location. Security also incorporates user authentication, which can be related to another important issue: billing. Both security and roaming must be based on the kind of service a user is accessing. The required QoS must be maintained when a user roams from one system to other. Besides maintaining the QoS, it should be possible to know what kind of service a particular system or service provider at a particular location can provide. Work on integration of WLANs and WPANs must also be done. The biggest technical challenge here will be enabling the coexistence of the two devices as both of them work on the same frequency band. FWA is a technology that should be watched as it develops. Depending on its market penetration and the development of standards, it should also be integrated with other technologies.

Another area of research for next generation communications will be in the field of personal networks (PNs) [27]. PNs provide a virtual space to users that spans a variety of infrastructure technologies and ad hoc networks. In other words PNs provide a personal distributed environment where people interact with various embedded or invisible computers not only in their vicinity, but potentially anywhere. Figure 1.9 portrays the concept of PNs. Several technical challenges arise with PNs besides interworking between different technologies, including security, self-organization, service discovery, and resource discovery [27].

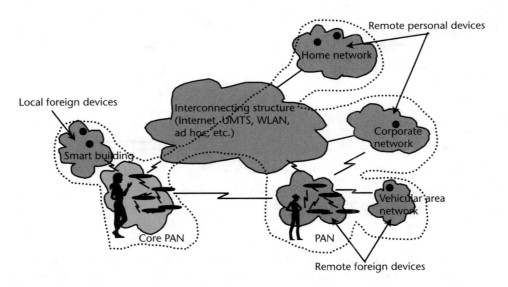

Figure 1.9 PN [27].

1.2   Orthogonal Frequency-Division Multiplexing

Over the past few years, there has been increasing emphasis on extending the serv-ices available on wired public telecommunications networks to mobile/movable nonwired telecommunications users. At present, in addition to voice services, only low-bit-rate data services are available to mobile users. However, demands for wire-less broadband multimedia communication systems (WBMCS) are anticipated within both the public and private sectors. Wired networks are cannot support extension to wireless mobile networks because mobile radio channels are more con-taminated than wired data-transmission channels. We also cannot preserve the high QoS required in wired networks [2].

The mobile radio channel is characterized by multipath reception: the signal offered to the receiver contains not only a direct line-of-sight (LOS) radio wave, but also a large number of reflected radio waves that arrive at the receiver at different times. Delayed signals are the result of reflections from terrain features such as trees, hills, mountains, vehicles, or buildings. These reflected, delayed waves interfere with the direct wave and cause intersymbol interference (ISI), which in turn causes significant degradation of network performance. A wireless network should be designed to minimize adverse effects.

To create broadband multimedia mobile communication systems, it is necessary to use high-bit-rate transmission of at least several megabits per second. However, if digital data is transmitted at the rate of several megabits per second, the delay time of the delayed waves is greater than 1 symbol time. Using adaptive equalization techniques at the receiver is one method for equalizing these signals. There are prac-tical difficulties in operating this equalization at several megabits per second with compact, low-cost hardware.

To overcome such a multipath-fading environment with low complexity and to achieve WBMCS, this chapter presents an overview of the orthogonal frequency-division multiplexing (OFDM) transmission scheme. OFDM is one of the applica-tions of a parallel-data-transmission scheme, which reduces the influence of multipath fading and makes complex equalizers unnecessary.

1.2.1   History of OFDM

OFDM is a special case of multicarrier transmission, where a single data stream is transmitted over a number of lower-rate subcarriers (SCs). It is worth mention-ing here that OFDM can be seen as either a modulation technique or a multiplex-ing technique. One of the main reasons to use OFDM is to increase robustness against frequency-selective fading or narrowband interference. In a single-carrier system, a single fade or interferer can cause the entire link to fail, but in a multicarrier system, only a small percentage of the SCs will be affected. Error-correction coding can then be used to correct for the few erroneous SCs. The concept of using parallel-data transmission and frequency-division multiplex-ing (FDM) was developed in the mid-1960s [28, 29]. Some early development is traced back to the 1950s [30]. A U.S. patent was filed and issued in January 1970 [31].

In a classical parallel-data system, the total signal frequency band is divided into N nonoverlapping frequency subchannels. Each subchannel is modulated with a separate symbol, and then the N subchannels are frequency multiplexed. It seems good to avoid spectral overlap of channels to eliminate interchannel interference. However, this leads to inefficient use of the available spectrum. To cope with the inefficiency, the ideas proposed in the mid-1960s were to use parallel data and FDM with overlapping subchannels, in which each, carrying a signaling rate b, is spaced b apart in frequency to avoid the use of high-speed equalization and to combat impulsive noise and multipath distortion, as well as to use the available bandwidth fully.

Figure 1.10 illustrates the difference between the conventional nonoverlapping multicarrier technique and the overlapping multicarrier modulation technique. By using the overlapping multicarrier modulation technique, we save almost 50% of bandwidth. To realize this technique, however, we need to reduce cross talk between SCs, which means that we want orthogonality between the different modulated carriers.

The word "orthogonal" indicates that there is a precise mathematical relationship between the frequencies of the carriers in the system. In a normal FDM system, many carriers are spaced apart in such a way that the signals can be received using conventional filters and demodulators. In such receivers, guard bands are introduced between the different carriers and in the frequency domain, which results in a lowering of spectrum efficiency.

It is possible, however, to arrange the carriers in an OFDM signal so that the sidebands of the individual carriers overlap and the signals are still received without adjacent carrier interference. To do this the carriers must be mathematically orthogonal. The receiver acts as a bank of demodulators, translating each carrier down to dc, with the resulting signal integrated over a symbol period to recover the raw data. If the other carriers all beat down the frequencies that, in the time domain, have a whole number of cycles in the symbol period T, then the integration process results in zero contribution from all of these other carriers. Thus, the carriers are

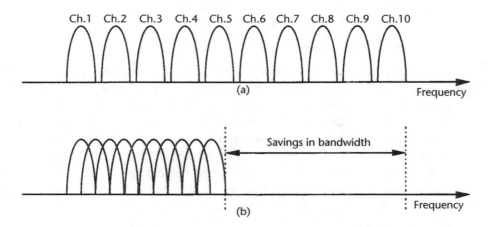

Figure 1.10 Concept of the OFDM signal: (a) conventional multicarrier technique, and (b) orthogonal multicarrier modulation technique.

linearly independent (i.e., orthogonal) if the carrier spacing is a multiple of $1/T$. Chapter 4 presents in detail the basic principle of OFDM.

Much of the research focuses on the highly efficient multicarrier transmission scheme based on "orthogonal frequency" carriers. In 1971, Weinstein and Ebert [32] applied the discrete Fourier transform (DFT) to parallel-data-transmission systems as part of the modulation and demodulation process. Figure 1.11(a) shows the spectrum of the individual data of the subchannel. The OFDM signal, multiplexed in the individual spectra with a frequency spacing b equal to the transmission speed of each SC, is shown in Figure 1.11(b). Figure 1.11 shows that at the center frequency of each SC, there is no cross talk from other channels. Therefore, if we use DFT at the receiver and calculate correlation values with the center of frequency of each SC, we recover the transmitted data with no cross talk. In addition, using the DFT-based multicarrier technique, FDM is achieved not by bandpass filtering but by baseband processing.

Moreover, to eliminate the banks of SC oscillators and coherent demodulators required by FDM, completely digital implementations could be built around special-purpose hardware performing the fast Fourier transform (FFT), which is an efficient implementation of the DFT. Recent advances in very-large-scale integration (VLSI) technology make high-speed, large-size FFT chips commercially affordable. Using this method, both transmitter and receiver are implemented using efficient FFT techniques that reduce the number of operations from N^2 in DFT to $N\log N$ [33].

In the 1960s, the OFDM technique was used in several high-frequency military systems such as KINEPLEX [30], ANDEFT [34], and KATHRYN [35]. For example, the variable-rate data modem in KATHRYN was built for the high-frequency band. It used up to 34 parallel low-rate phase-modulated channels with a spacing of 82 Hz.

In the 1980s, OFDM was studied for high-speed modems, digital mobile communications, and high-density recording. One of the systems realized the OFDM techniques for multiplexed quadrature amplitude modulation (QAM) using DFT [36]; also, by using pilot tone, stabilizing carrier and clock frequency control and trellis coding could also be implemented [37]. Moreover, various-speed modems were developed for telephone networks [38].

In the 1990s, OFDM was exploited for wideband data communications over mobile radio FM channels, high-bit-rate digital subscriber lines (HDSL; 1.6 Mbps),

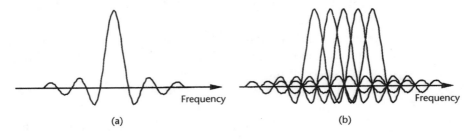

Figure 1.11   Spectra of (a) an OFDM subchannel, and (b) an OFDM signal.

asymmetric digital subscriber lines (ADSL; up to 6 Mbps), **very-high**-speed digital subscriber lines (VDSL; 100 Mbps), digital audio broadcasting (DAB), and high-definition television (HDTV) terrestrial broadcasting [39–46].

The OFDM transmission scheme has the following key advantages:

- OFDM is an efficient way to deal with multipath; for a given delay spread, the implementation complexity is significantly lower than that of a single-carrier system with an equalizer.
- In relatively slow time-varying channels, it is possible to enhance capacity significantly by adapting the data rate per SC according to the signal-to-noise ratio (SNR) of that particular SC.
- OFDM is robust against narrowband interference because such interference affects only a small percentage of the SCs.
- OFDM makes single-frequency networks possible, which is especially attractive for broadcasting applications.

On the other hand, OFDM also has some drawbacks compared with single-carrier modulation:

- OFDM is more sensitive to frequency offset and phase noise.
- OFDM has a relatively large peak-to-average-power ratio, which tends to reduce the power efficiency of the radio frequency (RF) amplifier.

1.3 Concluding Remarks

Multicarrier techniques, including OFDM-based wireless systems, will provide the solution for future-generation wireless communications. The following provides some of the justification:

1. Multicarrier techniques can combat hostile frequency-selective fading encountered in mobile communications. The robustness against frequency-selective fading is very attractive, especially for high-speed data transmission.
2. OFDM scheme has matured well through research and development for high-rate WLANs and terrestrial DVB. We have developed a lot of know-how for OFDM.
3. Combining OFDM with CDMA yields synergistic effects, such as enhanced robustness against frequency-selective fading and high scalability in possible data-transmission rates.

Figure 1.12 shows the advantages of multicarrier techniques.

The real challenge for the future can be explained by (1.1) to achieve IP-based wireless multimedia communications:

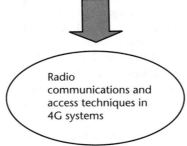

Figure 1.12   Advantages of multicarrier techniques for 4G systems.

$$E \propto m.c^4 \tag{1.1}$$

where E is evolution of wireless communications, m is multimedia communications, and c is consumer electronics, computer technology, communications technology, and contents. Figure 1.13 illustrates the clue to the evolution/revolution of wireless IP-based multimedia communications.

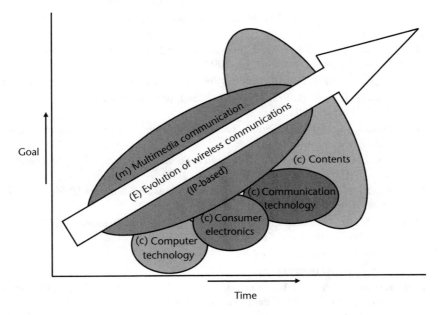

Figure 1.13   Evolution/revolution of wireless IP-based multimedia communications.

References

[1] Prasad, R., "60-GHz Systems and Applications," *Second Annual Workshop on 60-GHz WLAN Systems and Technologies,* Kungsbacka, Sweden, May 15–16, 2001.

[2] Prasad, R., "Wireless Broadband Communication Systems," *IEEE Communications Magazine,* Vol. 35, January 1997, p. 16.

[3] Correia, L. M., and R. Prasad, "An Overview of Wireless Broadband Communications," *IEEE Communications Magazine,* Vol. 35, January 1997, pp. 28–33.

[4] Prasad, R., and L. M. Correia, "Wireless Broadband Multimedia Communication," *Proceedings of the International Wireless and Telecommunications Symposium,* Shah Alam, Malaysia, May 14–16, 1997, Vol. 2, pp. 55–70.

[5] Prasad, R., and L. M. Correia, "An Overview of Wireless Broadband Communications," *Proc. MoMuC'97,* Seoul, Korea, September/October 1997, pp. 17–31.

[6] Prasad, R., *CDMA for Wireless Personal Communications,* Norwood, MA: Artech House, 1996.

[7] Prasad, R., *Universal Wireless Personal Communications,* Norwood, MA: Artech House, 1998.

[8] Ojanperä, T., and R. Prasad, (eds.), *Wideband CDMA for Third Generation Mobile Communications,* Norwood, MA: Artech House, 1998.

[9] van Nee, R., and R. Prasad, *OFDM for Wireless Multimedia Communications,* Norwood, MA: Artech House, 1999.

[10] Prasad, R., W. Mohr, and W. Konhäuser, (eds.), *Third Generation Mobile Communication Systems,* Norwood, MA: Artech House, 2000.

[11] Ojanperä, T., and R. Prasad, (eds.), *WCDMA: Towards IP Mobility and Mobile Internet,* Norwood, MA: Artech House, 2000.

[12] Prasad, R., (ed.), *Towards a Global 3G System: Advanced Mobile Communications in Europe, Volume 1,* Norwood, MA: Artech House, 2001.

[13] Prasad, R., (ed.), *Towards a Global 3G System: Advanced Mobile Communications in Europe, Volume 2,* Norwood, MA: Artech House, 2001.

[14] Farserotu, J., and R. Prasad, *IP/ATM Mobile Satellite Networks,* Norwood, MA: Artech House, 2001.

[15] Harada, H., and R. Prasad, *Simulation and Software Radio for Mobile Communications,* Norwood, MA: Artech House, 2002.

[16] Dixit, S., and R. Prasad, *Wireless Internet,* Norwood, MA: Artech House, 2002.

[17] Munoz, L., and R. Prasad, *WLANs and WPANs towards 4G Wireless,* Norwood, MA: Artech House, 2003.

[18] Prasad, R., and M. Ruggieri, *Technology Trends in Wireless Communication,* Norwood, MA: Artech House, 2003.

[19] Hara, S., and R. Prasad, *Multicarrier Techniques for 4G Mobile Communications,* Norwood, MA: Artech House, 2003.

[20] Prasad, N. R., and A. Prasad, *WLAN Systems and Wireless IP for Next Generation Communication,* Norwood, MA: Artech House, 2002.

[21] Prasad, A., "Wireless LANs: Protocols, Security and Deployment," Ph.D. thesis, Delft University of Technology, the Netherlands, 2003.

[22] http://www.nttdocomo.com/i/.

[23] http://www.wapforum.org.

[24] Kamerman, A., and A. R. Prasad, "IEEE 802.11 and HIPERLAN/2 Performance and Applications," *ECWT 2000,* Paris, France, October 2–6, 2000.

[25]   Farserotu, J., et al., "Scalable, Hybrid Optical-RF Wireless Communication System for Broadband and Multimedia Service to Fixed and Mobile Users," Invited Paper, *International Journal on Wireless Personal Communications*, Kluwer Academic Publishers, Vol. 24, No. 2, 2003, pp. 327–329.

[26]   Lauridsen, O. M., and A. R. Prasad, "User Needs for Services in UMTS," *International Journal on Wireless Personal Communications*, Vol. 22, No. 2, August 2002, pp. 187–197.

[27]   Niemegeers, I. G., and S. M. Heemstra de Groot, "Research Issues in Ad-Hoc Distributed Personal Networks," *International Journal on Wireless Personal Communications*, Vol. 26, Iss. 2–3, September 2003, pp. 149–167.

[28]   Chang, R. W., "Synthesis of Band Limited Orthogonal Signals for Multichannel Data Transmission," *Bell Syst. Tech. J.*, Vol. 45, December 1996, pp. 1775–1796.

[29]   Salzberg, B. R., "Performance of an Efficient Parallel Data Transmission System," *IEEE Trans. Communications*, Vol. COM-15, December 1967, pp. 805–813.

[30]   Mosier, R. R., and R. G. Clabaugh, "Kineplex, a Bandwidth Efficient Binary Transmission System," *AIEE Trans.*, Vol. 76, January 1958, pp. 723–728.

[31]   "Orthogonal Frequency Division Multiplexing," U.S. Patent No. 3,488,4555, filed November 14, 1966, issued January 6, 1970.

[32]   Weinstein, S. B., and P. M. Ebert, "Data Transmission by Frequency Division Multiplexing Using the Discrete Fourier Transform," *IEEE Trans. Communications*, Vol. COM-19, October 1971, pp. 628–634.

[33]   Zou, W. Y., and Y. Wu, "COFDM: An Overview," *IEEE Trans. Broadcasting*, Vol. 41, No. 1, March 1995, pp. 1–8.

[34]   Porter, G. C., "Error Distribution and Diversity Performance of a Frequency Differential PSK HF Modem," *IEEE Trans. Communications*, Vol. COM-16, August 1968, pp. 567–575.

[35]   Zimmerman, M. S., and A. L. Kirsch, "The AN/GSC-10 (KATHRYN) Variable Rate Data Modem for HF Radio," *IEEE Trans. Communications*, Vol. COM-15, April 1967, pp. 197–205.

[36]   Hirosaki, B., "An Orthogonally Multiplexed QAM System Using the Discrete Fourier Transform," *IEEE Trans. Communications*, Vol. COM-29, July 1981, pp. 982–989.

[37]   Hirosaki, B., "A 19.2 Kbits Voice Band Data Modem Based on Orthogonality Multiplexed QAM Techniques," *Proc. of IEEE ICC'85*, 1985, pp. 21.1.1–5.

[38]   Keasler, W. E., and D. L. Bitzer, "High Speed Modem Suitable for Operating with a Switched Network," U.S. Patent No. 4,206,320, June 1980.

[39]   Chow, P. S., J. C. Tu, and J. M. Cioffi, "Performance Evaluation of a Multichannel Transceiver System for ADSL and VHDSL Services," *IEEE J. Selected Area*, Vol. SAC-9, No. 6, August 1991, pp. 909–919.

[40]   Chow, P. S., J. C. Tu, and J. M. Cioffi, "A Discrete Multitone Transceiver System for HDSL Applications," *IEEE J. Selected Areas in Communication*, Vol. SAC-9, No. 6, August 1991, pp. 909–919.

[41]   Paiement, R. V., *Evaluation of Single Carrier and Multicarrier Modulation Techniques for Digital ATV Terrestrial Broadcasting*, CRC Report No. CRC-RP-004, Ottawa, Canada, December 1994.

[42]   Sari, H., G. Karma, and I. Jeanclaude, "Transmission Techniques for Digital Terrestrial TV Broadcasting," *IEEE Comm. Mag.*, Vol. 33, February 1995, pp. 100–109.

[43]   Oppenheim, A. V., and R. W. Schaffer, *Discrete-Time Signal Processing*, Upper Saddle River, NJ: Prentice Hall, 1989.

[44]   Hara, S., et al., "Transmission Performance Analysis of Multi-Carrier Modulation in Frequency Selective Fast Rayleigh Fading Channel," in *Wireless Personal Communications*, 2nd ed., New York: Kluwer Academic Publishers, 1996, pp. 335–356.

[45]   Rappaport, T. S., *Wireless Communications: Principles and Practice*, Upper Saddle River, NJ: Prentice Hall, 1996.

[46]   Steel, R., (ed.), *Mobile Radio Communications*, Piscataway, NJ: IEEE Press, 1994.

CHAPTER 2

WLANs

2.1 Introduction

The past decade has shown major changes in the types of communications services provided to users and the infrastructure needed to support them. Besides the present-day telephony, Internet access, applications with remote servers, video on demand, and interactive multimedia are just a few examples of such services. Internet access is the service that has captured the biggest market and enjoys maximum penetration; this is shown in Figure 2.1 for year and number of users. Wireline communications networks providing these services are mostly known as wide area networks (WANs) and LANs.

The overall market demand is basically for connectivity, mobility, and performance. Wireline services can provide connectivity and performance, but not mobility together with connectivity; this market demand is depicted in Figure 2.2. Wireless communications provide the solution to the requirements of mobility with connectivity. Thus, together with the growth of the Internet, there has been tremendous growth in the field of wireless communications. This has also been due to other inherent benefits of wireless, namely decreased wiring complexity, increased flexibility, and ease of installation. The main reason behind the growth of wireless has

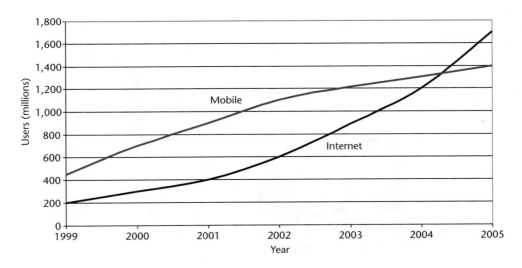

Figure 2.1 Growth in wireless and Internet.

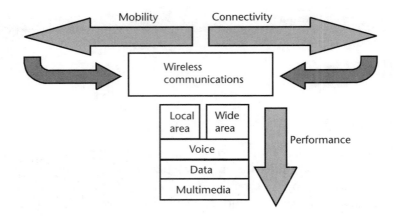

Figure 2.2 Market trend.

been WWANs or mobile technologies based on 2G/2.5G standards like Global System for Mobile Communications (GSM) and Personal Digital Cellular (PDC). These technologies mainly provide voice services and some data services at low data rates. 3G systems provide higher data rates with a maximum throughput of 2 Mbps (see Figure 2.3).

WLANs, on the other hand, provide connectivity, lower mobility, and much higher performance in terms of achievable data rate. They are mainly extension of LANs providing high-speed data services with lower mobility. Complementary to WLANs are WPANs, which provide wireless data networking within a short range

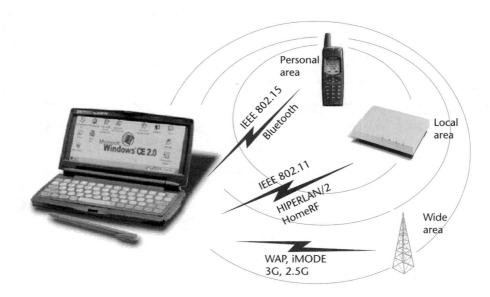

Figure 2.3 Wide area, local area, and personal area wireless technologies.

(~10m) at data rates of about 1 Mbps. A summary of WWANs, WLANs, and WPANs standards is given in Figure 2.3 [1–55].

WLANs provide a new forum of access technology in the LAN world. The new access technology fulfils several practical requirements (increased mobility, flexibility), but several technical problems remain unsolved. The problems of WLANs are tackled by researchers throughout the world.

2.1.1   WLANs in a Nutshell

WLANs operate mostly using either radio or infrared techniques. Each approach has it own attributes, which satisfy different connectivity requirements. The majority of these devices are capable of transmitting information across distances of up to several hundred meters in an open environment. Figure 2.4 shows a WLAN interfacing with a wired network. The WLAN consists of a wireless network interface card, known as station (STA), and a wireless bridge referred to as an access point (AP). The AP interfaces the wireless network with the wired network (e.g., Ethernet LAN) [10–16, 28].

The most widely used WLANs use radio waves at the frequency band of 2.4 GHz, also known as the industrial, scientific and medical (ISM) band. The worldwide availability of the ISM band, shown in Figure 2.5 and Appendix 2A, has made unlicensed spectrum available and promoted significant interest in the design of WLANs. An advantage of radio waves is that they can provide connectivity for non-LOS situations. A disadvantage of radio waves is that the electromagnetic propagation may cause interference with equipment working at the same frequency. Because radio waves propagate through the walls, security might also be a problem. Further details of ISM band standards is given in Appendix 2B.

WLANs based on radio waves usually use spread spectrum technology. Spread spectrum spreads the signal power over a wide band of frequencies, which makes

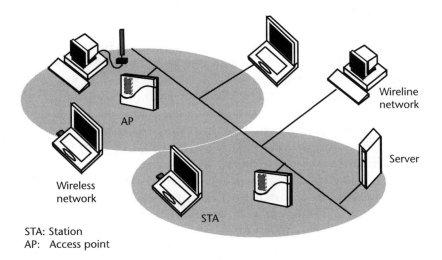

Figure 2.4   A WLAN.

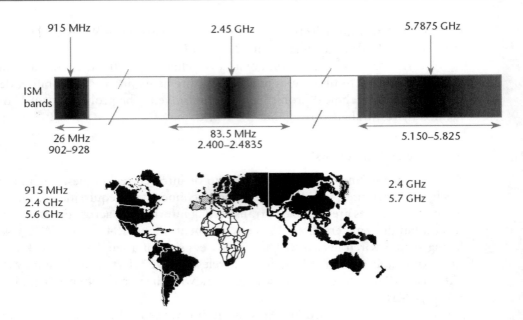

Figure 2.5 Worldwide availability of ISM bands.

the data much less susceptible to electrical noise than when using conventional radio modulation techniques. Spread spectrum modulators use one of two methods to spread the signal over a wider spectrum: frequency-hopping spread spectrum (FHSS) or direct-sequence spread spectrum (DSSS) [29].

FHSS works very much as the name implies. It takes the data signal and modulates it with a carrier signal that hops from frequency to frequency as a function of time over a wide band of frequencies. On the other hand, DSSS combines a data signal at a sender with a higher data-rate bit sequence, thus spreading the signal over the whole frequency band [28, 29]. Infrared LANs provide an alternative to radio wave–based WLANs. Although infrared has its benefits, it is not suitable for many mobile applications due to its LOS requirements [28].

The first WLAN products appeared on the market around 1990, although the concept of WLANs had been around for some years. The next generation of WLAN products were implemented on PCMCIA cards (also called PC cards) used in laptop computers and portable devices. In recent years several WLAN standards have come into being. IEEE 802.11–based WLAN [10, 11, 28] was the first and remains the most prominent in the field. IEEE 802.11 has different physical layers (PHYs) working in the 2.4- and 5-GHz bands.

Other WLAN standards are (1) HomeRF [13], now considered dead, dedicated to the home market based on FHSS, and (2) High Performance LAN Type 2 (HIPERLAN/2) [12, 17, 21, 28], which works in the 5-GHz band using the OFDM technique.

The exponential growth of the Internet and wireless has brought about tremendous changes in LAN technology. WLAN technology is becoming more and more important. Although WLAN been around since the early 1990s, the market has just started opening and the technology is still ripening. WLAN use is envisaged for several environments like the home, office, and hotspots to name a few (see Figure 2.6).

2.1.2   IEEE 802.11, HIPERLAN/2, and MMAC WLAN Standards

Since the beginning of the 1990s, WLANs for the 900-MHz, 2.4-GHz, and 5-GHz ISM bands have been available, based on a range of proprietary techniques. In June 1997, the IEEE approved an international interoperability standard [10]. The standard specifies both MAC procedures and three different PHYs. There are two radio-based PHYs using the 2.4-GHz band. The third PHY uses infrared light. All PHYs support a data rate of 1 Mbps and optionally 2 Mbps. The 2.4-GHz frequency band is available for license-exempt use in Europe, the United States, and Japan. Table 2.1 lists the available frequency bands and the restrictions for devices that use these bands for communications.

Table 2.1   International 2.4-GHz ISM Bands

Location	Regulatory Range	Maximum Output Power
North America	2.400–2.4835 GHz	1,000 mW
Europe	2.400–2.4835 GHz	100 mW (EIRP*)
Japan	2.471–2.497 GHz	10 mW

* EIRP = effective isotropic radiated power

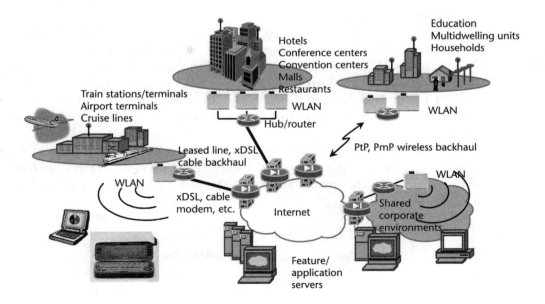

Figure 2.6   Envisaged WLAN usage environments.

User demand for higher bit rates and the international availability of the 2.4-GHz band has spurred the development of a higher-speed extension to the 802.11 standard. In July 1998, a proposal was selected for standardization, which describes a PHY providing a basic rate of 11 Mbps and a fallback rate of 5.5 Mbps. This PHY can be seen as a fourth option, to be used in conjunction with the MAC that is already standardized. Practical products, however, are expected to support both the high-speed 11- and 5.5-Mbps modes as well as the 1- and 2-Mbps modes.

A second IEEE 802.11 WG has moved on to standardize yet another PHY option, which offers higher bit rates in the 5.2-GHz band. This development was motivated by the adoption, in January 1997, by the U.S. Federal Communications Commission, of an amendment to Part 15 of its rules. The amendment makes available 300 MHz of spectrum in the 5.2-GHz band, intended for use by a new category of unlicensed equipment called Unlicensed National Information Infrastructure (UNII) devices [51]. Table 2.2 lists the frequency bands and corresponding power restrictions. Notice that the maximum permitted output power depends on the emission bandwidth; for a bandwidth of 20 MHz, you are allowed to transmit at the maximum power levels listed in the middle column of Table 2.2. For a bandwidth smaller than 20 MHz, the power limit reduces to the value specified in the right column.

Like the IEEE 802.11 standard, the European ETSI HIPERLAN Type 1 standard [52] specifies both MAC and PHY. Unlike IEEE 802.11, however, no HIPERLAN Type 1–compliant products are available in the market place. A newly formed ETSI WG called Broadband Radio Access Networks (BRAN) is now working on extensions to the HIPERLAN standard. Three extensions are under development: HIPERLAN/2, a wireless indoor LAN with a QoS provision; HiperLink, a wireless indoor backbone; and HiperAccess, an outdoor, fixed wireless network providing access to a wired infrastructure.

In Japan, equipment manufacturers, service providers, and the Ministry of Post and Telecommunications are cooperating in the Multimedia Mobile Access Communication (MMAC) project to define new wireless standards similar to those of IEEE 802.11 and ETSI BRAN. Additionally, MMAC is also looking into the possibilities for ultra-high-speed wireless indoor LANs supporting large-volume data transmission at speeds of up to 156 Mbps using frequencies in the 30–300-GHz band.

In July 1998, the IEEE 802.11 standardization group decided to select OFDM as the basis for its new 5-GHz standard, targeting a range of data rates from 6 up to 54 Mbps [53, 54]. This new standard is the first to use OFDM in packet-based

Table 2.2   U.S. 5.2-GHz UNII Band

Location	Maximum Output Power	Minimum Of
5.150–5.250 GHz	50 mW	$4 \text{ dBm} + 10\log_{10}B*$
5.250–5.350 GHz	250 mW	$11 \text{ dBm} + 10\log_{10}B$
5.725–5.825 GHz	1,000 mW	$17 \text{ dBm} + 10\log_{10}B$

B is the –26-dB emission bandwidth in megahertz.

communications; the use of OFDM until now has been limited to continuous transmission systems like DAB and digital video broadcasting (DVB). Following the IEEE 802.11 decision, ETSI BRAN and MMAC also adopted OFDM for their PHY standards. The three bodies have worked in close cooperation since then to make sure that differences among the various standards are kept to a minimum, thereby enabling the manufacturing of equipment that can be used worldwide.

The focus of this section is on the PHY side. In the case of the IEEE 802.11 standard, the MAC layer for the higher data rates remains the same as for the currently supported 1- and 2-Mbps rates. A description of this MAC can be found in [54].

2.1.2.1 OFDM Parameters

Table 2.3 lists the main parameters of the draft OFDM standard. A key parameter that largely determines the choice of the other parameters is the guard interval (GI) of 800 ns. This GI provides robustness to rms delay spreads of up to several hundred nanoseconds, depending on the coding rate and modulation used. In practice, this means that the modulation is robust enough to be used in any indoor environment, including large factory buildings. It can also be used in outdoor environments, although directional antennas may be needed to reduce the delay spread to an acceptable amount and increase the range.

To limit the relative amount of power and time spent on the guard time to 1 dB, the symbol duration chosen is 4 μs. This also determines the SC spacing at 312.5 kHz, which is the inverse of the symbol duration minus the guard time. By using 48 data SCs, uncoded data rates of 12 to 72 Mbps can be achieved by using variable modulation types from binary phase shift keying (BPSK) to 64-QAM. In addition to the 48 data SCs, each OFDM symbol contains an additional four pilot SCs, which can be used to track the residual carrier frequency offset that remains after an initial frequency correction during the training phase of the packet.

To correct for SCs in deep fades, forward error correction (FEC) across the SCs is used with variable coding rates, giving coded data rates from 6 to 54 Mbps.

Table 2.3 Main Parameters of the OFDM Standard

Data rate	6, 9, 12, 18, 24, 36, 48, 54 Mbps
Modulation	BPSK, QPSK, 16-QAM, 64-QAM
Coding rate	1/2, 2/3, 3/4
Number of SCs	52
Number of pilots	4
OFDM symbol duration	4 μs
Guard interval	800 ns
SC spacing	312.5 kHz
−3-dB bandwidth	16.56 MHz
Channel spacing	20 MHz

Convolutional coding is used with the industry standard rate 1/2, constraint length 7 code with generator polynomials (133,171). Higher coding rates of 2/3 and 3/4 are obtained by puncturing the rate 1/2 code. The 2/3 rate is used together with 64-QAM only to obtain a data rate of 48 Mbps. The 1/2 rate is used with BPSK, QPSK, and 16-QAM to give rates of 6, 12, and 24 Mbps, respectively. Finally, the 3/4 rate is used with BPSK, quadrature phase shift keying (QPSK), 16-QAM, and 64-QAM to give rates of 9, 18, 36, and 54 Mbps, respectively.

2.1.2.2   Differences Between IEEE 802.11, HIPERLAN/2, and MMAC

The main differences between IEEE 802.11 and HIPERLAN/2—which is standard-ized by ETSI BRAN [55]—are in the MAC. IEEE 802.11 uses a distributed MAC based on Carrier Sense Multiple Access with Collision Avoidance (CSMA/CA), while HIPERLAN/2 uses a centralized and scheduled MAC based on wireless asyn-chronous transfer mode (ATM). MMAC supports both of these MACs. As far as the PHY is concerned, there are a few relatively minor differences between IEEE 802.11 and HIPERLAN/2, which are summarized next.

HIPERLAN uses different training sequences. The long training symbol is the same as for IEEE 802.11, but the preceding sequence of short training symbols is dif-ferent. A downlink transmission starts with 10 short symbols as IEEE 802.11, but the first 5 symbols are different in order to detect the start of the downlink frame. The rest of the packets in the downlink frame do not use short symbols, only the long training symbol. Uplink packets may use 5 or 10 identical short symbols, with the last short symbol being inverted.

HIPERLAN uses extra puncturing to accommodate the tail bits to keep an inte-ger number of OFDM symbols in 54-byte packets. This extra operation punctures 12 bits out of the first 156 bits of a packet.

In the case of 16-QAM, HIPERLAN uses a coding rate of 9/16 instead of 1/2, giving a bit rate of 27 instead of 24 Mbps, to get an integer number of OFDM sym-bols for packets of 54 bytes. The rate 9/16 is made by puncturing 2 out of every 18 encoded bits.

Both IEEE 802.11 and HIPERLAN scramble the input data with a length 127 pseudorandom sequence, but the initialization is different. IEEE 802.11 initializes with seven random bits, which are inserted as the first seven bits of each packet. In HIPERLAN, the scrambler is initialized by {1, 1, 1}, plus the first four bits of the broadcast channel at the beginning of a MAC frame. The initialization is identical for all packets in a MAC frame. HIPERLAN devices have to support power control in the range of −15 to 30 dBm with a step size of 3 dB.

Dynamic frequency selection is mandatory in Europe over a range of at least 330 MHz for indoor products and 255 MHz (upper band only) for outdoor products. This means that indoor products have to support a frequency range from 5.15 to at least 5.6 GHz, covering the entire lower band and a part of the European upper band. Dynamic frequency selection was included to avoid the need for frequency planning and to provide coexistence with radar systems that operate in the upper part of the European 5-GHz band.

2.2 MAC in WLAN Standards

The MAC protocols form the basis of efficient use of a channel, be it wireline or wireless. When numerous users desire to transmit on a channel at the same time, conflicts occur, so there must be procedures on how the available channel capacity is allocated. These procedures constitute the MAC protocol rules each user has to follow in accessing the common channel [30]. The channel thus becomes a shared resource whose allocation is critical to the proper functioning of the network. With the boom of WLANs, an efficient MAC has become a must.

To design an appropriate MAC protocol, one has to understand the wireless network under discussion [30–32]. The first things that should be understood are a system's duplexing scheme and the network architecture. A MAC protocol is dependent on these two issues.

Duplexing refers to mechanisms for wireless devices to send and receive. There are two duplexing methods: time-based or frequency-based. Sending and receiving data in same frequency at different time periods is known as time-division duplexing (TDD), while sending and receiving data at the same time at different frequencies is known as frequency-division duplexing (FDD).

A wireless network can be distributed or centralized. In distributed networks each device accesses the medium individually and transmits the data without any central control. Distributed network architectures require the same frequency and thus makes use of TDD. IEEE 802.11 is an example of distributed network architecture. On the other hand, a centralized network architecture has one network element that controls the communication of various devices. Such network architectures can make use of both TDD and FDD. HIPERLAN/2 is an example of centralized network architecture.

This section discusses the MAC protocols in IEEE 802.11 [33, 34] and HIPERLAN/2 [35]. As IEEE 802.11 is the most commonly used WLAN, it is explained in more detail.

2.2.1 IEEE 802.11

IEEE 802.11 was standardized to satisfy the needs of wireless data networking. In CSMA/CA, the MAC protocol adopted by IEEE 802.11 [3, 10], the basic channel access method is random backoff CSMA with a MAC-level acknowledgment. A CSMA protocol requires the STA to listen before talking. In this protocol only one user can access the medium at a time, while the system is mostly used for low-data-rate applications (Internet access, e-mail).

IEEE 802.11 basic medium access behavior allows interoperability between compatible PHYs through the use of the CSMA/CA protocol and a random backoff time following a busy medium condition. In addition, all traffic uses immediate positive acknowledgment (ACK), where the sender schedules a retransmission if no ACK is received. The IEEE 802.11 CSMA/CA protocol is designed to reduce the collision probability between multiple stations accessing the medium at the point where collisions would most likely occur. Collisions are most likely to happen just after the medium becomes free, that is, just after busy medium conditions, because multiple stations would have been waiting for the medium to become available again. Therefore, a random backoff arrangement is used to resolve medium-contention conflicts. The IEEE 802.11 MAC also describes the way beacon frames are sent by the AP at regular intervals (like 100 ms) to enable stations to monitor the presence of the AP. The MAC also gives a set of management frames that allow a station to scan actively for other APs on any available channel. Based on this information the station may decide on the best-suited AP. In addition, the 802.11 MAC defines special functional behavior for the fragmentation of packets, medium reservation via request-to-send/clear-to-send (RTS/CTS) polling interactions, and point coordination (for time-bounded services) [33].

The MAC sublayer is responsible for channel allocation procedures, protocol data unit (PDU) addressing, frame formatting, error checking, and fragmentation and reassembly. The transmission medium can operate in the contention mode exclusively, requiring all stations to contend for access to the channel for each packet transmitted. The medium can also alternate between the contention mode, known as the contention period (CP), and a contention-free period (CFP). During the CFP, medium usage is controlled (or mediated) by the AP, thereby eliminating the need for stations to contend for channel access. IEEE 802.11 supports three different types of frames: management, control, and data. Management frames are used for station association and disassociation with the AP, timing and synchronization, and authentication and deauthentication. Control frames are used for handshaking during the CP, for positive acknowledgments during the CP, and to end the CFP. Data frames are used for the transmission of data during the CP and CFP and can be combined with polling and acknowledgments during the CFP.

As the contention-free mode is not used, this section will discuss the contention mode of IEEE 802.11 MAC, which is also known as the distributed coordination function (DCF). The RTS/CTS mechanism of IEEE 802.11 is not discussed in this chapter. The IEEE 802.11 MAC discussed here is the original MAC and not IEEE 802.11e or i, which present work on QoS and security, respectively.

2.2.1.1 DCF

The DCF is the fundamental access method used to support asynchronous data transfer on a best-effort basis. As identified in the IEEE 802.11 specification [3, 10], all stations must support the DCF. The DCF operates solely in the ad hoc network and either operates solely or coexists with the point coordination function (PCF) in an infrastructure network. Figure 2.7 depicts the MAC architecture and shows that the DCF sits directly on top of the PHY and supports contention services. Contention services imply that each station with a packet queued for transmission must contend for access to the channel and, once the packet is transmitted, must recontend for access to the channel for all subsequent frames. Contention services promote fair access to the channel for all stations [33].

The DCF is based on CSMA/CA. In IEEE 802.11, carrier sensing is performed at both the air interface, referred to as physical carrier sensing, and at the MAC sublayer, referred to as virtual carrier sensing. Physical carrier sensing detects the

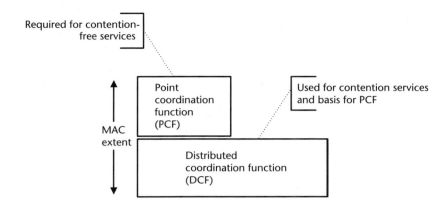

Figure 2.7   MAC architecture.

presence of other IEEE 802.11 WLAN users by analyzing all detected packets and also detects activity in the channel via relative signal strength from other sources.

A source station performs virtual carrier sensing by sending packet duration information in the header of RTS, CTS, and data frames. A packet is a complete data unit that is passed from the MAC sublayer to the PHY. The packet contains header information, payload, and a 32-bit cyclic redundancy check (CRC). The duration field indicates the amount of time (in microseconds) after the end of the present frame that the channel will be utilized to complete the successful transmission of the data or management frame. Stations in the basic service set (BSS) use the information in the duration field to adjust their network allocation vector (NAV), which indicates the amount of time that must elapse until the current transmission session is complete and the channel can be sampled again for idle status. The channel is marked busy if either the physical or virtual carrier sensing mechanisms indicate the channel is busy.

Priority access to the wireless medium is controlled through the use of interframe space (IFS) time intervals between the transmissions of frames. The IFS intervals are mandatory periods of idle time on the transmission medium. Three IFS intervals (see Figure 2.8) are specified in the standard: short IFS (SIFS), point coordination function IFS (PIFS), and DCF-IFS (DIFS). The SIFS interval is the smallest IFS, followed by PIFS, then DIFS. Stations only required to wait a SIFS period have priority access over those stations required to wait a PIFS or DIFS period before transmitting; therefore, SIFS has the highest-priority access to the communications medium. For the basic access method, when a station senses the channel is idle, the station waits for a DIFS period and samples the channel again. If the channel is still idle, the station transmits an MAC protocol data unit (MPDU). The receiving station calculates the checksum and determines whether the packet was received correctly. Upon receipt of a correct packet, the receiving station waits a SIFS interval and transmits a positive ACK frame back to the source station, indicating that the transmission was successful. Figure 2.9 is a timing diagram illustrating the successful transmission of a data frame. When the data frame is transmitted, the duration field of the frame is used to let all stations in the BSS know how long the medium

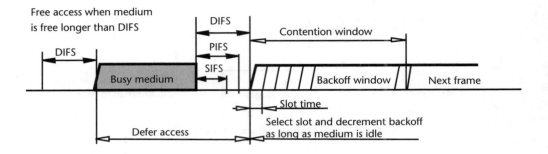

Figure 2.8   IEEE 802.11 IFS.

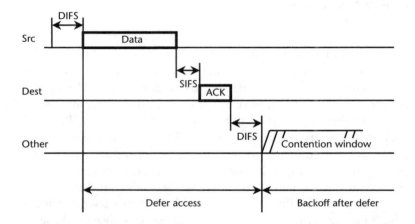

Figure 2.9   Transmission of an MPDU without RTS/CTS.

will be busy. All stations hearing the data frame adjust their NAVs based on the duration field value, which includes the SIFS interval and the ACK following the data frame.

The collision avoidance portion of CSMA/CA is performed through a random backoff procedure. If a station with a frame to transmit initially senses that the channel is busy, then the station waits until the channel becomes idle for a DIFS period, then computes a random backoff time. For IEEE 802.11, time is slotted in time periods that correspond to a Slot_Time. The Slot_Time used in IEEE 802.11 is much smaller than an MPDU and is used to define the IFS intervals and determine the backoff time for stations in the CP. The Slot_Time is different for each PHY implementation. The random backoff time is an integer value that corresponds to a number of time slots. Initially, the station computes a backoff time in the range of zero to seven. Once the medium becomes idle after a DIFS period, stations decrement their backoff timer until the medium becomes busy again or the timer reaches zero. If the timer has not reached zero and the medium becomes busy, the station freezes its timer. When the timer is finally decremented to zero, the station transmits

its frame. If two or more stations decrement to zero at the same time, a collision will occur, leading to missing ACKs, and each station will have to generate a new back-off time in the range 0 to 63 for 802.11b or 0 to 31 for 802.11a, multiplied by the Slot_Time period. The generated backoff time corresponds to a uniform distributed integer multiple of Slot_Time periods. For the next retransmission attempt, the backoff time grows to 0 to 127 for 802.11b and 0 to 63 for 802.11a, and so on, with a maximum range of 0 to 1,023. The idle period after a DIFS period is referred to as the contention window (CW). The advantage of this channel access method is that it promotes fairness among stations, but its weakness is that it probably could not support time-bound services. Fairness is maintained because each station must recontend for the channel after every transmission of an MPDU. All stations have equal probability of gaining access to the channel after each DIFS interval. Time-bounded services typically support applications such as packetized voice or video that must be maintained with a specified minimum delay. With DCF, there is no mechanism to guarantee minimum delay to stations supporting time-bounded services.

2.2.2   HIPERLAN/2

The MAC in HIPERLAN/2 is a part of the data link control (DLC) layer together with other functions like error control (EC). This section provides a brief description of the MAC layer and frames of HIPERLAN/2 [35].

2.2.2.1   MAC Layer

The MAC scheme of HIPERLAN/2 is based on a central controller, which is located at the AP. The core task of the central controller is to determine the direction of information flow between the controller and the terminal at any point. A MAC frame consists of control and data blocks. The central controller decides which terminal or group of terminals is allowed to transmit in a slot of the frame. The medium access scheme is classified as load-adaptive time-division multiple access (TDMA). Each user shall be assigned zero, one, or several slots in a frame. In general, the number of slots assigned to an individual user varies from frame to frame and depends on the actual bandwidth request of the terminal. The uplink and downlink packets are sent on the same frequency channel in a TDD mode.

Random-access slots are provided to allow STAs to get associated with the controller. In this "bootstrap phase" data is transmitted in a contention-based mode, collisions may occur. Therefore, a collision resolution algorithm is applied.

Uplink signaling of resource describes the state of the input queues of a STA to the central controller. The AP collects these requests from all associated STAs and uses this data to schedule the uplink access times. The results of the scheduling process are signaled via the frame control channel; that is, a description of the exact frame structure and slot allocation is contained in the frame control channel. This control data is valid for the ongoing frame. Further tasks include (1) multiplexing and demultiplexing of logical channels, (2) service requesting and service granting, and (3) means for MAC.

2.2.2.2 MAC Frames

The MAC frame structure (Figure 2.10) comprises time slots for broadcast control, frame control, access feedback control, and data transmission in downlink, uplink, and directlink phases, which are allocated dynamically depending on the need for transmission resources. An STA first has to request capacity from the AP in order to send its data. This can be done in the random-access channel, where contention for the same time slot is allowed.

Downlink, uplink and directlink phases consist of two types of PDUs: long PDUs and short PDUs. The long PDUs (Figure 2.11) have a size of 54 bytes and contain control or user data. The payload is 49.5 bytes, and the remaining 4.5 bytes are used for the PDU type (2 bits), a sequence number (10 bits), and CRC (CRC-24 bits). Long PDUs are referred to as the long transport channel. Short PDUs contain only control data and have a size of nine bytes. They may contain resource requests, automatic repeat request (ARQ) messages, and the like, and they are referred to as the short transport channel.

Traffic from multiple connections to or from one STA can be multiplexed onto one PDU train, which contains long and short PDUs. A physical burst is composed of the PDU train payload and a preamble and is the unit to be transmitted via the PHY (see Figure 2.12).

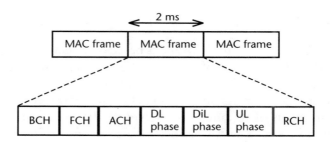

Figure 2.10   The HIPERLAN/2 MAC frame.

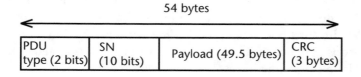

Figure 2.11   Format of the long PDUs.

Figure 2.12   Format of the PDU train.

2.3   QoS over WLANs

QoS is becoming an increasingly important element of any communications system. In the simplest sense, QoS means providing a consistent, predictable, data delivery service, or in other words, satisfying the customer application requirements. Providing QoS means providing real-time (e.g., voice) as well as nonreal-time services. Voice communication is the primary form of service required by mankind. The public telephone network and the equipment that makes it possible are taken for granted in most parts of the world. The availability of a telephone and access to a low-cost, high-quality worldwide network is considered essential to a modern society (telephones are even expected to work when the power is off).

Support for voice communications using the IP, which is usually just called Voice over IP (VoIP), has become especially attractive given the low-cost, flat-rate pricing of the public Internet. VoIP can be defined as the ability to make telephone calls (i.e., to do everything that can be done today with the Public Switched Telecommunications/Telephone Network or PSTN) over IP-based data networks with suitable QoS. This is desirable because the cost/benefit ratio is far superior to that of the PSTN. Equipment producers see VoIP as a new opportunity to innovate and compete. The challenge for them is turning this vision into reality by quickly developing new VoIP-enabled equipments that are capable of providing toll-quality service. For Internet service providers (ISPs) the possibility of introducing usage-based pricing and increasing traffic volume is very attractive. Both the ISPs and the network manufacturers face the challenge of developing and producing solutions that can provide the required voice quality. Users are seeking new types of integrated voice/data applications, as well as cost benefits.

As WLANs are extensions of IP to the wireless, it is necessary to have a Voice over WLAN (VoWLAN) protocol that fulfils the requirement. Figure 2.13 depicts a complete system for VoWLAN, IP to plain old telephone service (POTS). Successfully delivering VoWLAN presents a tremendous opportunity; however, implementing these products is not as straightforward a task as it may first appear.

This section presents the QoS mechanisms adopted in the present IEEE 802.11e (IEEE 802.11 standard for MAC enhancements for QoS) draft standard.

2.3.1   IEEE 802.11e

IEEE 802.11e provides MAC enhancements to support LAN applications with QoS requirements. The QoS enhancements are available to the QoS-enhanced stations (QSTAs) associated with a QoS-enhanced access point (QAP) in a QoS-enabled network. A subset of the QoS enhancements may be available for use between QSTAs. A QSTA may associate with a non-QoS AP in a non-QoS network [6, 7]. Non-QoS STAs may associate with a QAP. Figure 2.14 depicts the MAC architecture of IEEE 802.11e.

The enhancements that distinguish the QSTAs from non-QoS STAs and the QAPs from non-QoS APs comprise an integrated set of QoS-related formats and functions that are collectively termed the *QoS facility*. The quantity of certain QoS-specific mechanisms may vary among QoS implementations, as well as between the

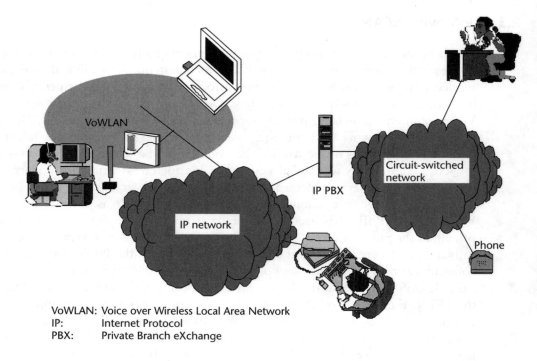

VoWLAN: Voice over Wireless Local Area Network
IP: Internet Protocol
PBX: Private Branch eXchange

Figure 2.13   VoWLAN, IP to POTS.

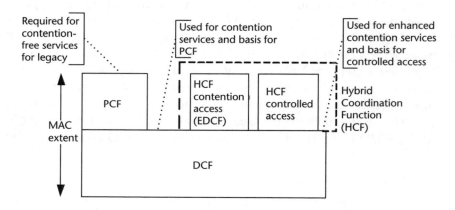

Figure 2.14   IEEE 802.11e MAC architecture.

QSTAs and the QAPs [7]. However, all service primitives, frame formats, coordination function and frame exchange rules, and management interface functions defined as part of the QoS facility are mandatory, with the exception of the group acknowledgement function defined in [7], which is an option separate from the core QoS facilities. Its presence is indicated by QSTAs separately from the core QoS facility.

The IEEE 802.11e standard provides two mechanisms for the support of applications with QoS requirements.

The first mechanism, designated as the enhanced DCF (EDCF), is based on the differentiating priorities at which the traffic is to be delivered. This differentiation is achieved through varying the amount of time a station would sense the channel to be idle, the length of the CW during a backoff, or the duration for which a station may transmit once it has the channel access.

The second mechanism allows for the reservation of transmission opportunities with the hybrid coordinator (HC). A QSTA based on its requirements requests the HC for transmission opportunities, both for its own transmissions as well as for transmissions from the HC to itself. The HC, based on an admission control policy, either accepts or rejects the request. If the request is accepted, it schedules transmission opportunities for the QSTA. For transmissions from the STA, the HC polls a QSTA based on the parameters supplied by the QSTA at the time of its request. For transmissions to the QSTA, the HC queues the frames and delivers them periodically, again based on the parameters supplied by the QSTA. This mechanism is expected to be used for applications such as voice and video, which may need a periodic service from the HC. This mechanism is a hybrid of several proposals studied by the standardization committee.

2.3.2   Interframe Spacing

The time interval between the frames is called the IFS. A STA determines that the medium is idle through the use of the carrier sense function for the interval specified. Five different IFSs are defined to provide priority levels for access to the wireless media; they are listed in order, from the shortest to the longest, except for the arbitration IFS (AIFS). Figure 2.15 shows some of these relationships. The different IFSs are independent of the STA data rate.

- SIFS;
- PIFS;

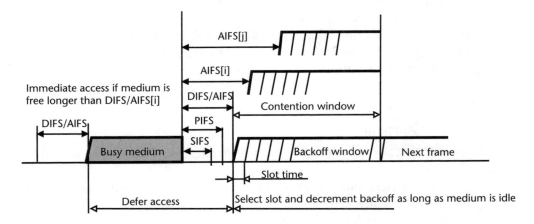

Figure 2.15   Interframe spacing for enhanced MAC.

- DIFS;
- AIFS (used by the QoS facility);
- Extended IFS.

The AIFS is to be used by QSTAs to transmit data and management frames. A QSTA using the EDCF is allowed a transmit opportunity (TxOP) for a particular traffic class (TC) if its carrier sense mechanism determines that the medium is idle at the TxAIFS(TC) slot boundary following a correctly received frame and the backoff time for that TC has expired.

2.3.3   Other QoS-Related Developments

IEEE 802.11e is also looking into admission control and scheduling, which will of course complete the picture for QoS support. Besides this the IEEE 802.11 standard is also looking into overlapping cell issues, which are very important for QoS, especially for a system with very few, unlicensed, nonoverlapping channels. Another ongoing work is IEEE 802.11f, which looks at the IAPP (approved as a standard in July 2003) [15]. Transferring context information from one AP to another will help in fast and seamless handover.

2.4   Security in IEEE 802.11

Barely a decade ago computers communicating by wireline medium replaced filing cabinets with strong combination locks. Communicating through a wireline medium meant the end (almost) of physical means of protecting sensitive information and a major change in the requirements for information security. Now, the communicating medium is becoming wireless. WLAN usage is exploding, and the once envisaged usage environments have already been reached: the markets of academia, enterprises, and the public. Security is an issue that can cause a major setback to the growth of WLANs [36–50].

This section discusses security in the current IEEE 802.11 standard and its security issues [43]. The section also discusses the current draft of IEEE 802.11i, the IEEE 802.11 security enhancements standard, and the IEEE 802.11f standard for fast mobility (IEEE 802.11 uses the word roaming, which is the same as handover in mobile communications).

2.4.1   Current IEEE 802.11

2.4.1.1   Authentication

IEEE 802.11 defines two subtypes of authentication service: open system and shared key [43].

Open-system authentication is the simplest of the available authentication algorithms. Essentially, it is a null authentication algorithm. Any STA that requests authentication with this algorithm may become authenticated if the recipient station is set to open-system authentication.

Shared-key authentication supports authentication of the STA as either a member of those who know a shared secret key or a member of those who do not. IEEE 802.11 shared-key authentication accomplishes this without the need to transmit the secret key in the clear, requiring the use of the wired equivalent privacy (WEP) mechanism. Therefore, this authentication scheme is only available if the WEP option is implemented. The required secret shared key is presumed to have been delivered to participating STAs via a secure channel that is independent of IEEE 802.11. During the shared-key-authentication exchange, both the challenge and the encrypted challenge are transmitted. This facilitates unauthorized discovery of the pseudorandom number (PRN) sequence for the key/initialization vector (IV) pair used for the exchange. Therefore, the same key/IV pair for subsequent frames should not be used. Figure 2.16 shows the shared-key-authentication process.

2.4.1.2   WEP

The WEP algorithm is a sort of electronic codebook in which a block of plaintext is bitwise XORed with a pseudorandom key sequence of equal length. The key sequence is generated by the WEP algorithm.

Referring to Figure 2.17 and viewing from the left to the right, the encipherment begins with a secret key that has been distributed to the cooperating STAs by an external key-management service. WEP is a symmetric algorithm in which the same key is used for encipherment and decipherment.

The secret key is concatenated with an IV, and the resulting seed is an input to the pseudorandom number generator (PRNG). The PRNG outputs a key sequence k of pseudorandom octets equal in length to the number of data octets that are to be transmitted in the MPDU, plus four [since the key sequence is used to protect the integrity check value (ICV) as well as the data]. Two processes are applied to the plaintext MPDU. To protect against unauthorized data modification, an integrity algorithm operates on the plaintext MPDU to produce an ICV. Encipherment is then accomplished by mathematically combining the key sequence with the

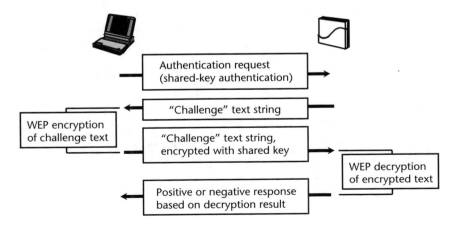

Figure 2.16   Shared-key authentication.

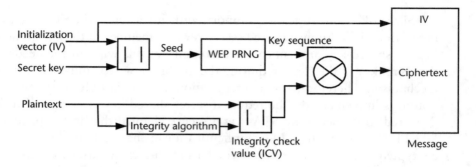

Figure 2.17   WEP encipherment block diagram.

plaintext concatenated with the ICV. The output of the process is a message containing the IV and ciphertext.

In Figure 2.18, from left to right, the decipherment begins with the arrival of a message. The IV of the incoming message shall be used to generate the key sequence necessary to decipher the incoming message. Combining the ciphertext with the proper key sequence yields the original plaintext and the ICV. Correct decipherment is verified by performing the integrity check algorithm on the recovered plaintext and comparing the output ICV' to the ICV transmitted with the message. If ICV' is not equal to ICV, the received MPDU is in error and an error indication is sent to the MAC management.

2.4.1.3   IEEE 802.11 Security Issues

The following security issues of IEEE 802.11 are known at the time of this writing:

- *Shared-key authentication:* Shared-key authentication suffers a known plaintext attack; recovering the pseudorandom string by XORing the plaintext and ciphertext of challenge, which can be eavesdropped from the air. Then the pseudorandom string can be used in a new authentication, even though the "shared secret" is not recovered.

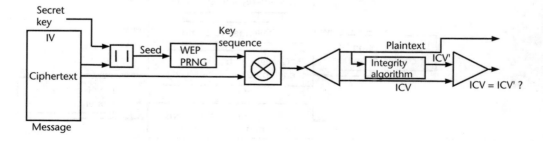

Figure 2.18   WEP decipherment block diagram.

- *Mutual authentication:* WEP provides no mutual authentication between the station and the AP; that is, the AP can authenticate the station, but not vice versa.
- *Key management:* There is no real key management in WEP, but two methods of using WEP keys are provided. The AP and the stations share the use of the four (default) keys. The compromising of each of the nodes means a compromise of the wireless network. A key mappings table is used at the AP. By this method, each unique MAC address can have a separate key. The size of a key mappings table should be at least 10 entries according to the 802.11 specification; however, it is likely chip-set dependent. The use of a separate key for each user mitigates the known cryptographic attacks, but requires more effort in manual key management. Since key distribution is not defined in WEP and can be done only manually, many of the organizations deploying wireless networks use either a permanent fixed cryptographic variable, or key, or no encryption at all.
- *Other problems:* Because WLAN APs are usually connected to intranets, which are protected by firewalls, a compromise of the WLAN can result in a serious exposure of the intranets. Using the same key for authentication and encryption increases the possibility of compromise.

For details on security issues, see [44].

2.4.2   IEEE 802.11i and IEEE 802.11f

This section provides an overview of IEEE 802.11i, security enhancements of IEEE 802.11, and IEEE 802.11f, IAPP. IEEE 802.11f was recently accepted as a standard. A short description of IEEE 802.1X is also given as it plays an important role in future WLAN security solutions.

2.4.2.1   IEEE 802.11i

IEEE 802.11i is a security enhancement standard currently in the draft stage.

As the IEEE 802.11i standard is not yet available, IEEE 802.11 vendors are providing security solutions to bridge the gap. These solutions started with an extended WEP key size, which was adopted by the standard, providing RADIUS and MAC address–based authentication, IEEE 802.1X port-based user authentication, and Advanced Encryption Standard (AES)–based encryption.

Seeing the market situation, WiFi, the IEEE 802.11 interoperability industry alliance, is introducing the Temporal Key Integrity Protocol (TKIP) as a simple, but secure, intermediary solution [11]. This solution is usually known as WiFi Protected Access (WPA) and is already available from some vendors. The WPA provides enhanced data encryption through TKIP, user authentication via IEEE 802.1X and EAP, and mutual authentication; for ease of transition, WiFi certified products are software upgradeable.

The following gives an overview of the current IEEE 802.11i draft.

The Robust Security Network

IEEE 802.11i defines a robust security network (RSN). An RSN provides a number of additional security features not present in the basic IEEE 802.11 architecture. These features include

- Enhanced authentication mechanisms for both the APs and the STAs;
- Key management algorithms;
- Dynamic, association-specific cryptographic keys;
- An enhanced data-encapsulation mechanism called Wireless Robust Authenticated Protocol (WRAP).

An RSN makes extensive use of protocols above the IEEE 802.11 MAC layer to provide authentication and key management. This allows the IEEE 802.11 standard both to take advantage of the work already done by other standardization bodies as well as to avoid duplicating functions at the MAC layer that are already performed at the higher layers. An RSN introduces several new components into the IEEE 802.11 architecture:

- *IEEE 802.1X port:* This is present on all STAs in an RSN and resides above the 802.11 MAC. All data traffic that flows through the RSN MAC also passes through the IEEE 802.1X port.
- *Authentication agent (AA):* This component resides on top of the IEEE 802.1X port at each STA and provide authentication and key management.
- *Authentication server (AS):* The AS is an entity that resides in the network and participates in the authentication of all STAs (including the APs). It may authenticate the elements of the RSN itself, or it may provide material that the RSN elements can use to authenticate each other.

As the IEEE 802.1X plays an important role in the IEEE 802.11i, it is explained separately in a later section.

Security Goals

An RSN does not directly provide the services. Instead, an RSN uses IEEE 802.1X to provide the access control and key distribution; the confidentiality is provided as a side effect of the key distribution. Some of the security goals of the IEEE 802.11i include the following:

- *Authentication:* An RSN-capable IEEE 802.11 network also supports upper-layer authentication based on IEEE 802.1X. Upper-layer authentication utilizes the protocols above the MAC layer to authenticate the STAs and the network with one another.
- *Deauthentication:* In an RSN using upper-layer authentication, the deauthentication may result in the disabling of the IEEE 802.1X-controlled port for the station.

- *Privacy:* IEEE 802.11i provides three cryptographic algorithms to protect the data traffic. Two are based on the RC4 algorithm defined by RSA, and the third is based on the AES. This standard refers to these as WEP, TKIP, and WRAP. A means is provided for the stations to select the algorithm to be used for a given association.

  WRAP, adopted by IEEE 802.11i, is based on the AES and the offset codebook (OCB).

- *Key distribution:* IEEE 802.11i supports two key distribution mechanisms: manual key distribution and automatic key distribution. Automatic key distribution is available only in an RSN that uses IEEE 802.1X to provide key distribution services. An RSN allows a number of authentication algorithms to be utilized. The standard does not specify a mandatory-to-implement upper-layer authentication protocol.

- *Data origin authentication:* This mechanism is available only to stations using WRAP and TKIP. Data origin authenticity is only applicable to unicast traffic.

- *Replay detection:* This mechanism is also available only to stations using the WRAP and the TKIP.

IEEE 802.1X

IEEE 802.1X is the standard for port-based network access control, which applies to IEEE 802.3 Ethernet, Token Ring, and WLAN [46]. Based on the Point-to-Point Protocol (PPP) Extensible Authentication Protocol (EAP) [47], IEEE 802.1X extends the EAP from PPP to the LAN applications. This standard defines EAP over LAN (EAPOL), a protocol that provides a framework for negotiating the authentication method. It defines no explicit authentication protocol itself and is extensible to many authentication protocols. It must be made clear that IEEE 802.1X is not an authentication protocol; nor does it guarantee a secure authentication algorithm for wireless applications.

Some of the terms used in the IEEE 802.1X and its relation with the WLAN are explained next.

- *IEEE 802.1X supplicant:* The entity at one end of the point-to-point LAN segment that is being authenticated; the software on the STA that implements the EAP;

- *Authenticator:* The entity that facilitates authentication of the entity attached to the other end of that link; the software on the AP that forwards the EAP control packets to the authentication server, enables/blocks the port, or uses received information;

- *Authentication server:* The entity that provides an authentication service to the authenticator; the Radius server, the Kerberos server, or the Diameter server; can be integrated into the AP.

Figure 2.19 shows the message sequence chart (MSC) of IEEE 802.1X. IEEE 802.1X has two ports. The data port at the authenticator is open until the authentication server authenticates the supplicant. Once the supplicant is authenticated, the

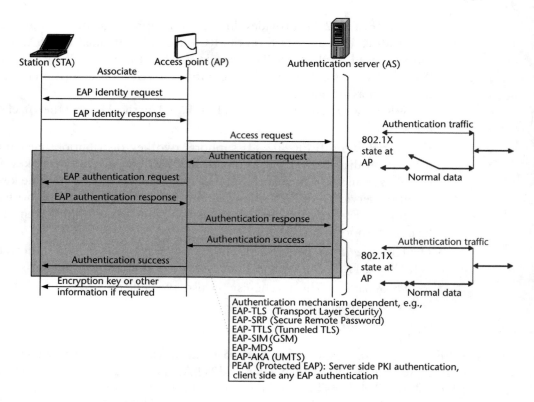

Figure 2.19   IEEE 802.1X EAPOL MSC.

data port is closed, and normal data communication can take place. IEEE 802.1X, together with the EAP, allows several different methods of authentication, some of which are also mentioned in the figure:

- EAP–Transport Layered Security;
- EAP–Secure Remote Password;
- EAP–Tunneled TLS;
- EAP–Subscriber Identity Module of GSM;
- EAP–Authentication and Key Agreement of Universal Mobile Telecommunications System (UMTS);
- EAP–Message Digest 5;
- Protected EAP.

2.4.2.2   IEEE 802.11f

IEEE 802.11f, IAPP, is a communication protocol used by one AP to communicate with other APs. It is a part of a communications system comprising APs, STAs, a backbone network, and the RADIUS infrastructure [48].

The RADIUS servers provide two functions:

1. Mapping the ID of an AP to its IP address;
2. Distribution of keys to the APs to allow the encryption of the communications between the APs.

The function of the IAPP is to facilitate the creation and maintenance of the wireless network, support the mobility of the STAs, and enable the APs to enforce the requirement of a single association for each STA at a given time.

One of the services the IAPP provides is proactive caching. Proactive caching is a method that supports fast roaming by caching the context of a STA in the APs to which the STA may roam. The next AP's are identified dynamically, that is, without management preconfiguration, by learning the identities of neighboring the APs.

References

[1]   Prasad, R., *Universal Wireless Personal Communications*, Norwood, MA: Artech House, 1998.

[2]   van Nee, R. D. J., and R. Prasad, *OFDM for Wireless Multimedia Communications*, Norwood, MA: Artech House, 2000.

[3]   Ojanpera, T., and R. Prasad, *Wideband CDMA for Third Generation Mobile Communications*, Norwood, MA: Artech House, 2000.

[4]   GSM MoU Web site at http://www.gsmworld.com.

[5]   ITU Web site at http://www.itu.int/imt.

[6]   The UMTS Forum Web site at http://www.umts-forum.org.

[7]   3GPP Web site at http://www.3gpp.org.

[8]   3GPP2 Web site at http://www.3gpp2.org.

[9]   *IEEE Personal Communications* Vol. 7, No. 2, April 2000.

[10] IEEE 802.11, Wireless LAN Medium Access Control (MAC) and Physical Layer (PHY) Specifications, November 1997.

[11] IEEE 802.11, Draft Supplement to Standard for Telecommunications and Information Exchange between Systems—LAN/MAN Specific Requirements—Part 11: Wireless MAC and PHY Specifications: High Speed Physical Layer in the 5-GHz Band, P802.11a/D6.0, May 1999.

[12] ETSI BRAN, HIPERLAN Type 2 Functional Specification Part 1—Physical Layer, DTS/BRAN030003-1, June 1999.

[13] HomeRF Web site at http://www.homerf.org.

[14] Bluetooth SIG Web site at http://www.bluetooth.com.

[15] *IEEE Personal Communications*, Vol. 7, No. 1, February 2000.

[16] IEEE 802.15, Part 15.1: Wireless Personal Area Network Medium Access Control (MAC) and Physical Layer (PHY) Specifications, May 2000.

[17] Kamerman, A., and A. R. Prasad, "IEEE 802.11 and HIPERLAN/2 Performance and Applications," *ECWT 2000*, Paris, France, October 2–6, 2000.

[18] Prasad, A. R., H. Moelard, and J. Kruys, "Security Architecture for Wireless LANs: Corporate and Public Environment," *VTC 2000*, Tokyo, Japan, May 15–18, 2000, pp. 283–287.

[19] Prasad, A., and A. Raji, "A Proposal for IEEE 802.11e Security," IEEE 802.11e, 00/178, July 2000.

[20]   Prasad, A. R., "Performance Comparison of Voice over IEEE 802.11 Schemes," *VTC 1999,* Amsterdam, the Netherlands, September 19–22, 1999, Fall 1999, pp. 2636–2640.

[21]   Prasad, N. R., et al., "A State-of-the-Art of HIPERLAN/2," *VTC 1999,* Amsterdam, the Netherlands, September 19–22, 1999, Fall 1999, pp. 2661–2666.

[22]   Prasad, A. R., "Optimization of Hybrid ARQ for IP Packet Transmission," *International Journal on Wireless Personal Communications,* Vol. 16, No. 3, March 2001, pp. 203–220.

[23]   Prasad, A. R., Y. Shinohara, and K. Seki, "Performance of Hybrid ARQ for IP Packet Transmission on Fading Channel," *IEEE Trans. on Vehicular Technology,* Vol. 48, No. 3, May 1999, pp. 900–910.

[24]   Prasad, A. R., and K. Seki, "Capacity Enhancement of Indoor Wireless Communication System with a Novel Channel Sharing Protocol," *ICPWC'97,* Mumbai, India, December 16–19, 1997, pp. 162–166.

[25]   Prasad, A. R., et al., "Wireless LANs Deployment in Practice," in *Wireless Network Deployments,* R. Ganesh and K. Pahelvan, (eds.), New York: Kluwer Publications, 2000.

[26]   Prasad, A. R., et al., "Performance Evaluation, System Design and Network Deployment of IEEE 802.11," *International Journal on Wireless Personal Communications,* Vol. 19, No. 1, October 2001, pp. 57–79.

[27]   Visser, M. A., and M. El Zarki, "Voice and Data Transmission over an 802.11 Network," *Proc. PIMRC'95,* Toronto, Canada, September 1995, pp. 648–652.

[28]   Prasad, N. R., and A. R. Prasad, (eds.), *WLAN Systems and Wireless IP for Next Generation Communications,* Norwood, MA: Artech House, 2002.

[29]   Prasad, R., *CDMA for Wireless Personal Communications,* Norwood, MA: Artech House, 1996.

[30]   van As, H. R., "Media Access Techniques: The Evolution Towards Terabit/s LANs and MANs," *Computer Networks and ISDN Systems,* Vol. 26, Nos. 6–8, 1994, pp. 603–656.

[31]   Chandra, A., V. Gummalla, and J. O. Limb, "Wireless Medium Access Control Protocols," *IEEE Communications Surveys,* Second Quarter 2000, at http://www.comsoc.org/pubs/surveys.

[32]   Rom, R., and M. Sidi, *Multiple Access Protocols Performance and Analysis,* New York: Springer-Verlag, 1990.

[33]   ISO/IEC 8802-11, ANSI/IEEE Std 802.11, First Edition 1999-00-00, Information Technology—Telecommunications and Information Exchange Between Systems—Local and Metropolitan Area Networks—Specific Requirements—Part 11: Wireless LAN Medium Access Control (MAC) and Physical Layer (PHY) Specifications.

[34]   Prasad, A. R., A. Kamerman, and H. Moelard, "IEEE 802.11 Standard," in *WLAN Systems and Wireless IP for Next Generation Communications,* N. R. Prasad and A. R. Prasad, (eds.), Norwood, MA: Artech House, 2002.

[35]   Prasad, N. R., and A. R. Prasad, "Wireless Networking and Internet Standards," in *WLAN Systems and Wireless IP for Next Generation Communications,* N. R. Prasad and A. R. Prasad, (eds.), Norwood, MA: Artech House, 2002.

[36]   Stallings, W., *Cryptography and Network Security: Principles and Practice,* Upper Saddle River, NJ: Prentice Hall, 1998.

[37]   Brederveld, L., N. R. Prasad, and A. R. Prasad, "IP Networking for Wireless Networks," in *WLAN Systems and Wireless IP for Next Generation Communications,* N. R. Prasad and A. R. Prasad, (eds.), Norwood, MA: Artech House, 2002.

[38]   Bishop, M., *Computer Security: Art and Science,* Reading, MA: Addison-Wesley, 2003.

[39]   Black, U., *Internet Security Protocols: Protecting IP Traffic,* Upper Saddle River, NJ: Prentice-Hall, 2000.

[40]   Borman, D., "Telnet Authentication: Kerberos Version 4," RFC 1411, January 1993.

[41] Kohl, J., and C. Neuman, "The Kerberos Network Authentication Service (V5)," RFC 1510, September 1993.

[42] Fox, A., and S. D. Gribble, "Security on the Move: Indirect Authentication Using Kerberos," *Proc. of 2nd ACM International Conference on Mobile Computing and Networking (MobiCom'96)*, Rye, NY, November 10–12, 1996.

[43] IEEE 802.11, Wireless LAN Medium Access Control (MAC) and Physical (PHY) Layer Specifications, ANSI/IEEE, 1999.

[44] Arbaugh, W. A., at http://www.cs.umd.edu/~waa/wireless.html.

[45] Rubens, A., et al., "Remote Authentication Dial In User Service (RADIUS)," RFC 2138, April 1997.

[46] IEEE Standard 802.1X-2001, IEEE Standard for Local and Metropolitan Area Networks—Port-Based Network Access Control, June 14, 2001.

[47] Blunk, L., and J. Vollbrecht, "PPP Extensible Authentication Protocol (EAP)," RFC 2284, March 1998.

[48] IEEE P802.11f, Draft Recommended Practice for Multi-Vendor Access Point Interoperability via an Inter-Access Point Protocol Across Distribution Systems Supporting IEEE 802.11 Operation, D5, January 2003.

[49] Prasad, A. R., H. Wang, and P. Schoo, *Network Operator's Security Requirements on Systems Beyond 3G*, WWRF #9, Zurich, Switzerland, July 1–2, 2003.

[50] Prasad, A. R., and P. Schoo, *IP Security for Beyond 3G Towards 4G*, WWRF #7, Eindhoven, the Netherlands, December 3–4, 2002.

[51] FCC, Amendment of the Commission's Rules to Provide for Operation of Unlicensed NII Devices in the 5-GHz Frequency Range, Memorandum Opinion and Order, ET Docket No. 96-102, June 24, 1998.

[52] ETSI, Radio Equipment and Systems, HIgh PErformance Radio Local Area Network (HIPERLAN) Type 1, European Telecommunication Standard, ETS 300-652, October 1996.

[53] Crow, B. P., et al., "IEEE 802.11 Wireless Local Area Networks," *IEEE Communications Magazine*, September 1997, pp. 116–126.

[54] Takanashi, H., and R. van Nee, *Merged Physical Layer Specification for the 5-GHz Band*, IEEE P802.11-98/72-r1, March 1998.

[55] ETSI, Broadband Radio Access Networks (BRAN); HIPERLAN Type 2 Technical Specification Part 1—Physical Layer, DTS/BRAN030003-1, October 1999.

Appendix 2A: ISM Bands

Location	Regulatory Range	Maximum Output Power	Standard
Europe	2,400–2,483.5 MHz 5,150–5,350 MHz 5,470–5,725 MHz	10 mW/MHz (maximum 100 mW) 200 mW 1,000 mW	IEEE 802.11b,g, HomeRF, WBFH, Bluetooth HIPERLAN/2 HIPERACCESS (FWA < 11 GHz) IEEE 802.11a
North America	2,400–2,483.5 MHz 5,150–5,250 MHz 5,250–5,350 MHz 5,725–5,825 MHz	1,000 mW 2.5 mW/MHz (maximum 50 mW) 12.5 mW/MHz (maximum 250 mW) 50 mW/MHz (maximum 1,000 mW)	IEEE 802.11b,g, HomeRF, WBFH, Bluetooth HIPERLAN Type 2 BWIF, IEEE 802.16 HUMAN IEEE 802.11a
Japan	2,400–2,497 MHz 5,150–5,250 MHz 4,900–5,000 MHz (until 2007) 5,030–5,091 MHz (from 2007)	10 mW/MHz (maximum 100 mW) Indoor 200 mW	IEEE 802.11b,g, HomeRF, WBFH, Bluetooth HIPERLAN Type 2 (MMAC HiSWAN) IEEE 802.11a (MMAC)

Appendix 2B: Comparison of WLAN and WPAN Standards

Standard	IEEE 802.11/b	IEEE 802.11a/g	HIPERLAN/2	IEEE 802.15 1.0 and Bluetooth	HomeRF
Mobile frequency range (MHz)	2,400–2,483 (North America/Europe) 2,470–2,499 (Japan)	a: 5,150–5,250 (Europe, North America, Japan) 5,250–5,350 (Europe, North America) 5,470–5,725 (Europe) 5,725–5,825 (North America) 4,900–5,000 (Japan) g: same as IEEE 802.11/b	Same as 802.11a	2,400–2483 (North America/Europe) 2,470–2499 (Japan)	2,400–2,483 (North America/Europe) 2,470–2,499 (Japan)
Multiple-access method	CSMA/CA (distributed and centralized)	CSMA/CA (distributed and centralized)	TDMA (centralized)	TDMA (centralized)	TDMA (distributed)/ CSMA (centralized)
Duplex method	TDD	TDD	TDD	FDD	TDD
Number of independent channels	FHSS:79 DSSS:3 to 5	a: 12 g: 3 to 5	12	FHSS: 79	FHSS: 79
Modulation	FHSS GFSK (0.5 Gaussian Filter) DSSS DBPSK (1 Mbps), DQSK (2 Mbps) b DSSS: CCK	a/g: OFDM 48 carriers 6 Mbps BPSK 1/2 9 Mbps BPSK 3/4 12 Mbps QPSK 1/2 18 Mbps QPSK 3/4 24 Mbps 16-QAM 1/2 36 Mbps 16-QAM 3/4 48 Mbps 64-QAM 2/3 54 Mbps 64-QAM 3/4 g PBCC and DSSS OFDM optional	OFDM 48 carriers 6 Mbps BPSK 1/2 9 Mbps BPSK 3/4 12 Mbps QPSK 1/2 18 Mbps QPSK 3/4 24 Mbps 16-QAM 1/2 36 Mbps 16-QAM 9/16 48 Mbps 64-QAM 3/4 54 Mbps 64-QAM 3/4	FHSS GFSK (0.5 Gaussian filter)	FHSS GFSK (0.5 Gaussian filter)
Channel bit rate (Mbps)	1 or 2 b 5.5 or 11	a/g: 6, 9, 12, 18, 24, 36, 48 and 54	6, 9, 12, 18, 24, 36, 48 and 54	1	1 or 2

CHAPTER 3

WPANs

3.1 Introduction

A WPAN is a networked collection of devices situated within a short range (10 meters is often a reference radius) [1–118]. The attribute "personal" stems from the fact that this collection of devices presumably belongs to an individual, forming his or her WPAN. Thus, the WPAN forms a wireless "bubble" around the person, referred to as the personal operating space (POS). Besides the connection among the personal devices that form the bubble, the WPAN should provide the user with a seamless, ad hoc connection to the compatible resources and APs that enter her POS. The WPAN bubble can expand and contract dynamically, depending on need; it may connect to the wall repeaters to access to the Internet or it can be dynamically stretched to include access to sensors and actuators.

The main design objectives of WPAN technology are low power consumption, operation in the unlicensed spectrum, low cost, and small package size. Low power consumption is a critical issue since the rate at which battery performance improves is fairly slow, compared to the explosive overall growth in wireless communications. Therefore, the wireless protocol itself should employ economic usage of battery energy. The WPAN systems use license-free wireless links because this is the only way to achieve ubiquitous connectivity without adverse impact on an existing wireless infrastructure. Finally, the low-cost single-chip solution sets the economic and ergonomic conditions for the wide spread of the WPAN technology.

3.1.1 Emergence of Personal Area Networking (The Person-Centered Concept)

The notion of the personal area network (PAN) is wider (and older) than the notion of the WPAN. The main goal of a PAN is, again, the networking of proximate devices, but it does not necessarily use the wireless medium. Instead, it may use the electric field with a human body as a conductor, magnetic field, and so forth. A good overview for PAN physical solutions can be found in [15]. Because the discussion here is chiefly restricted to WPAN, the terms *WPAN* and *PAN* will be used interchangeably. The cases where PAN is not WPAN will be explicitly noted.

The present notion of PAN came about as an accretion of several developments and tendencies. Some of them were strongly interrelated from the very beginning; nevertheless, all tendencies now tend to be merged in a unique conception. These factors led to the emergence of the PAN, which traced its independent evolutionary line afterward, defining its own application scenarios and motivating the appearance of new applications and services.

- *Bridging different wireless standards:* Today, we face a diverse set of wireless access technologies applied in wide area cellular networks (GSM, IS-95, IMT-2000), personal communication systems, and WLANs (802.11, HIPERLAN). Most of these systems, however, are still tailored toward a narrow and specific application scenario.

    Hence, there is a need for a single universal low-cost wireless communication system that offers a user-friendly and efficient way to access information using a variety of devices such as mobile PCs, mobile phones, PDAs, pagers, and digital cameras. Such a wireless solution would bring together all of these technologies applied in different sectors and at the same time provide a universal and ubiquitous connectivity solution between computing and communications devices.

- *Very high wireless data rates:* The user's need for bandwidth is increasing continuously. In fact, the need for higher data speeds drove the evolution of 2G wireless systems to the 3G UMTS. Further increasing data rates beyond UMTS requires the use of picocells.

    The low-power, picocellular nature of PANs implies high spatial capacity, defined as bits per second per square meter, or bps/m^2 [18]; that is, it enables a more efficient spatial reuse of the radio spectrum. The short-range wireless networks, such as PANs and WLANs, can support significantly higher data rates than the ones offered by the 3G wireless systems. Figure 3.1 depicts the mobility versus data-rate graph for existing and future wireless technologies.

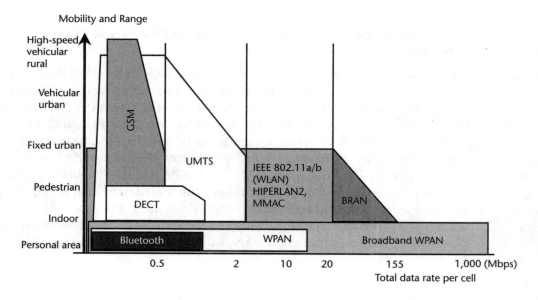

Figure 3.1 The settlement of existing and future wireless technologies.

- *Zimmermann's work on PAN:* Zimmerman [19] introduced the term *PAN* to denote a communication between proximal electronic devices by using the body as a conduit for information. This "body network" has been conceived as a support to the intelligent environment that the person "carries" herself. The motivation for such a network comes from the fact that there is a need for data exchange not only across large distances (which is conventionally referred to as communications), but also between people who are at a "conversational" distance.

  It may be concluded that Zimmerman's PAN is a concrete contribution to the trend of ubiquitous communications that largely inspires investigation of new classes of communication services with the attribute "personal."

- *Cable replacement:* This development did not bear the revolutionary flavor of Zimmerman's work, but occurred as a logical next step in the enhancement of computers and electronic appliances. Here we refer to the initiatives for developing cable-replacement or "last-meters" technologies instantiated through the specifications of the Infrared Data Association (IrDA), HomeRF, and Bluetooth WGs. Each of these technologies surpassed its initial target, offering far more flexibility to the electronic devices than mere cable replacement.

- *Ergonomic settlement of personal electronic devices:* This is closely related to cable replacement. The possibility of wireless interconnection of proximal devices motivates investigation of new computing structures directed toward calm technology [20]. For example, the PDA's keyboard can be a control interface to all other personal devices.

- *Ubiquity of Internet access:* The number of APs to the wired Internet has grown significantly. People need Internet access everywhere: at home, in enterprises, in public spaces. The WPAN will equip the individual with "wearable" Internet access.

- *Spectrum regulation issues:* The crowding of the radio spectrum has caused the development of wireless communication technologies that operate in the unlicensed spectrum. The paradigm of low-power spread spectrum radio systems with short radio coverage neatly fits into the troubled contest for utilization of the radio spectrum. And it is important to note that the extensive research activities on spread spectrum during the last decade promoted it to a mature and well-examined transmission technology, suitable for application.

- *Cheaper hardware:* Shrinking semiconductor costs and lower power consumption for signal processing make it feasible to build or upgrade personal computing devices with wireless communications capability.

3.1.1.1 Emergence of the Person-Centered Concept

There is a strong consensus that new technologies should be centered on the user, improving quality of life and adapting to the individual. The communications/computing technology tends toward "invisibility" and "calmness" [20]. The offered services tend to be pervasive, causing minimum distraction to the user with respect to their configuration and usage. The computing environment is becoming smarter and more responsive, with devices able to establish disposable, seamless connection to the required resource.

While the traditional communications paradigm aims to establish the communication link between devices, the focus now shifts to communication among persons and functional/data resources (here, the term *person* refers to all personal communication-enabled devices). The addresses of sources/destinations in communication links are determined either by the person who owns them, the service they are able to offer, or the resource's contents. This causes radical changes in design, for example, in addressing (content-based or capability-based), security, and so forth. As a consequence, new research fields are emerging, addressing different aspects of this problem. Examples are service portability and virtual home environments [21], concepts aimed at providing users with the same service experience independently of user interfaces, terminal capabilities, access network technologies, and network and service providers. Another important and related emerging area is pervasive computing targeting environments where networked computing devices are ubiquitous and even integrated with the human user [22].

The paradigm shift mentioned implies a different approach to the development of wireless communications. As the Wireless World Research Forum (WWRF) concluded [23], a purely technical vision for wireless development is not enough. In other words, the investigation of, for example, new network concepts or radio interfaces will not be sufficient to come to grips with the future. Rather, such a technical view must be broadened, complemented by the following:

- A person-centered approach, looking at the new ways users will interact with the wireless systems;
- New services and applications that become possible with the new technologies;
- New business models that may prevail in the future, overcoming the by now traditional user, server, provider hierarchy.

There is an essential difference in thinking about 4G wireless systems compared to the way 3G and other present wireless standards are produced. While the latter standards have been put in a technology-driven development process, early 4G philosophy is being approached from an application viewpoint, with an implied assumption that technology will follow to enable the realization of the application vision [16]. The essence is to provide a ubiquitous networking capability in which questions of data speeds are rendered irrelevant by the universal availability of more bandwidth than the vast majority of users would ever need.

The 4G wireless communications will tend toward the personal [1]. The user will no longer be "owned" by any operator: The users, or their smart agents, will select at each instant the best system available that is capable of providing the required performance. The selection will be made according to the user's profile, the type of data stream, and the traffic load in the available networks.

3.1.1.2 WPAN and the Person-Centered Concept

A fundamental concept behind WPAN systems asserts that any time two WPAN-equipped devices get within (approximately) 10m of one another, they can form a spontaneous, just-in-time, disposable connection. The purposes of such connections fall into three broad categories [17]:

1. *Leveraging device synergies:* This category is in accordance with the initial motives for the design of cable-replacement technologies. It refers to allowing devices to offer their capabilities to one another by establishing appropriate wireless data connections. Thus, for example, a laptop computer can serve as a configuration interface for a digital camera. In general, these synergistic connections will help solve the human-interface problems that arise from stuffing more and more complexity into smaller and smaller packages.

2. *Making queues obsolete:* People spend a lot of time waiting in a queue for access to the system needed to execute functions such as cash transactions or airline seat assignments. In most of these case, customers could serve themselves without waiting if they could obtain secure access to the same system via a handheld device like a PDA or cell phone equipped with WPAN capability.

3. *Grouping Internet users efficiently:* The motivation behind this usage of WPANs is the fact that most people spend the majority of their day within 10m of some kind of Internet port. The number of such places and hours per day spent in that state will only increase (e.g., in airports, hotels, shopping malls). Wherever densely packed users gather in small spaces, WPAN can offer data connections at much higher speeds, for many more users, with far longer battery life than that possible with cellular-based systems. Many expect the combination of wireless WPAN and wired Internet to become a fast-growing complement (and even alternative) to the next generation cellular systems for data, voice, audio, and video. A possible Internet access scenario of interconnected WPANs is shown on Figure 3.2.

Initially, WPAN technology offered an efficient cable replacement, easing the interconnection of communications, computing, and/or information storage devices. In the long term, this short-range wireless communication may largely influence the ways computing operations are conceived and performed. Thus, the ad hoc connectivity brought by the WPANs can motivate the design of the computing devices themselves, as well as the distribution of the computing tasks and capabilities over different devices. The prototypes of WPAN-distributed computing (to differentiate from the notion of distributed computing) can be seen in wearable computers. Wearable computers consist of head-mounted displays, microphones, earphones, processors, mass storage, and a diversity of control interfaces.

The POS, introduced by the WPAN, is a new concept in communications. The POS is tethered to an individual, moving along with him or her and enabling his or her personal devices to communicate in an ad hoc manner. It further allows the individual to communicate with other devices whose communication range intersects

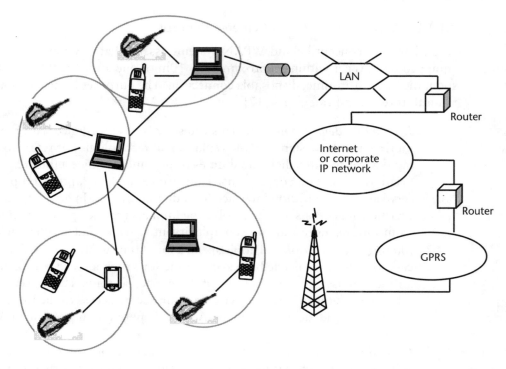

Figure 3.2   PAN scenario with four interconnected LANs, two of which have an Internet connection via a PAN–LAN AP and a GPRS/UMTS phone.

with his or her POS. Thus, the WPAN concept shifts the well-established communication paradigm. The POS plays the role of the universal access interface to the networked user.

3.2   Technical Challenges of a WPAN Technology

In the future WPAN should offer users seamless, autoconfigured connections to available resources. The (un)availability of the resources may be physical or protocol-dependent (e.g., compatibility, security). The WPAN will be implemented on low-cost chips embedded in wearable devices. The WPAN should be critically aware of the scarceness of energy in battery-powered devices. Therefore, low power consumption should be the essential requirement in technological design, with a higher priority than optimized spectral efficiency [1, 2]. Abstracting from the concrete applications and usage scenarios, WPAN should support the following two (interrelated) operations: Devices should be able to discover the service or information resource needed and to establish an ad hoc connection to the device that offers that service or contains that resource. Thus, ad hoc connectivity can be regarded as a generic capability and service discovery and selection as a generic application of personal area networking. Thus, for example, the procedure for gaining awareness of a proximate computing entity can be decomposed into ad hoc connection establishment and service discovery.

3.2.1   Ad Hoc Connectivity

Ad hoc connectivity refers to the capability of devices to establish mutual, on-demand, timely peer communication wireless links where an ad hoc communication network is created with little, if any, reconfiguration. This means that there is no fixed infrastructure to support network initiation, there is no central controller for the units to rely on for making interconnections, and there is no support for coordination of communications. The ad hoc network exists for and due to the cooperative functioning of the constituent wireless nodes.

Ad hoc networks have generated substantial interest in recent years [3–5]. Several distinctive features make the task of ad hoc network design highly challenging:

- *Distributed operation:* This feature is embedded in the definition of ad hoc networks. Networking functions, such as routing and security, must be supported by the cooperative operation of the participating wireless units.
- *Dynamic network topology:* The mobility of nodes and radio propagation conditions cause continuous changes in the network topology. This calls for employing particular network protocol functions for topology construction and maintenance.
- *Multihop communications:* Due to signal propagation characteristics and topology creation, ad hoc networks require the support of multihop communications; that is, the links between some nodes must be established by relaying via a third node or a group of nodes.
- *Restricted bandwidth and fluctuating link capacity:* The wireless environment possesses an inherent "hostility" toward the wireless communication systems, expressed in increased probability of bit and frame errors. The effects of these error rates are aggregated along the multihop paths, yielding higher link error rates.
- *Energy-constrained nodes:* The wireless units in the ad hoc network are (mainly) battery driven. Hence, operation with low-power consumption is a must for ad hoc network participants. This power optimization should be performed at all layers in the protocol stack and not only in terms of signal processing. An important way of achieving power efficiency is increased cross-layer interaction [6]. Power-efficiency may be taken into account in ad hoc topology maintenance [7] or the routing protocol [8].
- *Limited security:* The wireless links are more vulnerable to security threats than wired networks. In ad hoc wireless networks, the security problems are augmented due to the absence of a central network security administration.

The research into ad hoc networking has not been bound to any particular networking technology. Much work has been done in the formulation of distributed routing algorithms [9, 10]. The need for new routing approaches becomes immediately obvious in ad hoc networking because in truly ad hoc systems, there is no difference between radio units; that is, there are no distinctive base stations or terminals. Stated in terms of routing, the whole network is based on the idea that devices serve both as routers and hosts at the same time. Due to the dynamic and volatile nature of the topology defined by such nodes, the routing algorithms directly handle the mobility.

Security issues in ad hoc networking are discussed in [11, 12]. The main security problem is to create trusted relationships between the cryptographic public keys without the aid of a trusted third-party certification. In addition, there are other security issues specific to ad hoc networking, for example, the "battery exhaustion attack" as a special form of denial of service attack.

The PAN technology, especially Bluetooth, is probably the first commercial real-world network where ad hoc networking concepts fit very well and could help to create robust and flexible network connectivity. The ad hoc networking approach in investigating WPANs should result from the instantiation of the general ad hoc approach with WPAN-specific attributes. Thus, the restricted radio coverage area can be beneficial from the security viewpoint; the mobility of nodes and the topology dynamics are different in WPAN ad hoc networks compared to the wide area ad hoc networks.

Although the concepts of ad hoc networking are much broader than the WPAN scenarios (e.g., wide area ad hoc networks), the products that apply those concepts will most likely see light in the short, personal area range [13]. These products will mainly focus on facilitating communication between a user's personal devices, either for local traffic or as gateways to the Internet. The ad hoc network functionality will also enable the interconnection of different users' devices—for instance, to facilitate larger ad hoc WGs. The intrinsic ability to create generic, small-scale, ad hoc networks in portable devices represents an entirely new area for future ad hoc-based applications.

3.2.2   Service Discovery and Resource Selection

Service discovery is in fact a part of distributed computing. In its simplest variant, the service discovery enables a computing device to have access to the service that is available within its communication range. For example, a PDA finds a printer within its proximity, recognizes it as an available computer resource (provided that certain security conditions are satisfied), and uses it as if the printer were installed in the PDA's software. Generally stated, service discovery refers to the process of establishing an on-demand connection at higher protocol stack levels (ultimately with seamless connection at the application level) with hardware that is physically available. With service discovery, devices may automatically discover network services including their properties, and services may advertise their existence in a dynamic way. In a generalized case, the service sought might not be within physical communication range of the entity that initiates the service discovery (one-hop service discovery). Nevertheless, it should eventually be possible to access the service, provided that there is still network connection between the entity and the service. In this case, the network connection may be a multihop ad hoc connection, but it may also use proximate networking infrastructure (e.g., AP and wired network).

The best-known current service discovery protocols (SDPs) are Jini, the IETF Service Location Protocol (SLP), Microsoft's Universal Plug and Play (UPnP), and the Information Access Service (IAS) protocol, defined for short-range infrared communications. One of the first WPAN-motivated protocols of this type is the Bluetooth SDP. All of them use similar system architectures, but differ in their functionality, network transport protocol, and the possibility of code mobility. The Jini project [14] allows devices to create spontaneous networks when plugged into each other, that is, when two devices come within wireless proximity. The goal of the project is to eliminate device configuration and driver installation. Similar projects, such as JavaSpaces, Hive, Tspaces, and GinJo, are discussed in [15].

There are several basic issues to address in service discovery. Clearly, the service capability of an entity present at a certain layer of the protocol stack should be incorporated in its addressing so that it can "propagate through" the lower layers and be reachable by the service requester. A capability description refers to the common way of describing the functions required by the requester and the functions provided by other devices and applications. The SDP itself must be designed to operate in a heterogeneous network environment.

The research into service discovery obtains a new dimension with the introduction of personal area networking. The WPAN SDP should take into account the low processing power and storage capabilities of wearable devices. In an ideal case, a certain service should be discovered and become available to a wearable device belonging to someone's WPAN as soon as there is a relay device within the POS that can route the request to the service-containing entity. This calls for an integrated approach to link establishment, route finding, and service discovery initiation.

3.3   Enabling Technologies

There are several competing technologies or standards that provide wireless connectivity within a short range. The most notable among them are Bluetooth, IrDA, HomeRF, and WLAN. These technologies compete on certain fronts and are complementary in other areas. None of the existing technologies possesses the target adaptability of the envisioned WPAN technology, but this is a consequence of the context in which the specifications or standards were created. Although each has targeted slightly different applications and usage models, the premise behind all of these standards is to use some kind of underlying radio technology to enable the wireless transmission of data and to provide support for the formation of networks and managing various devices by means of high-level software.

However, the ubiquity and seamless connectivity of the WPAN should not be regarded only as a goal that each technology separately tends to achieve. The potential of each of the competing technologies to gain widespread use poses the problem of their coexistence as well as of their cooperation and interoperability in providing service This implies that the development of the WPAN technology should consist not only of a solution superior to existing solutions, but it should also offer to glue together the different technologies pervasive at the time the superior solution appears.

We first give brief descriptions of the relevant existing technologies, as well as of UWB as an emerging technology.

- *IrDA [34]:* This is an international organization that creates and promotes interoperable, low-cost, infrared data interconnection standards. IrDA has a set of protocols covering all layers of data transfer and, in addition, has some network management and interoperability designs. IrDA protocols have IrDA DATA as the vehicle for data delivery and IrDA CONTROL for sending the control information. In general, IrDA is used to provide wireless connectivity technologies for devices that would normally use cables for connectivity.

  IrDA is a point-to-point, narrow-angle (30° cone), ad hoc data transmission standard designed to operate over a distance of 0 to 1m and at speeds of 9,600 bps to 16 Mbps. Adapters now include the traditional upgrades to serial and parallel ports.

- *HomeRF [35]:* A subset of the International Telecommunication Union (ITU), HomeRF primarily works on the development of a standard for inexpensive RF voice and data communication.

  The HomeRF WG has also developed the Shared Wireless Access Protocol (SWAP). SWAP is an industry specification that permits PCs, peripherals, cordless telephones, and other devices to communicate voice and data without the use of cables. It uses a dual protocol stack: DECT for voice, and 802.11 packets for data. It is robust, reliable, and minimizes the impact of radio interference. Its target applications are home networking, as well as remote control and automation.

- *Bluetooth:* Bluetooth is a high-speed, low-power microwave wireless link technology designed to connect phones, laptops, PDAs and other portable equipment with little or no work on the part of the user. Unlike infrared, Bluetooth does not require LOS positioning of connected units. The technology uses modifications of existing WLAN techniques, but is most notable for its small size and low cost. Whenever Bluetooth-enabled devices come within range of each other, they instantly transfer address information and establish small networks between themselves without user involvement. To a large extent, Bluetooth has motivated the present WPAN attempts and conceptualizations; moreover, it constitutes the substance of the IEEE 802.15.1. WPAN standard.

- *IEEE 802.15 WPAN:* The 802.15 WPAN effort focuses on the development of consensus standards for PANs or short-distance wireless networks. These WPANs address wireless networking of portable and mobile computing devices such as PCs, PDAs, peripherals, cell phones, pagers, and consumer electronics, allowing these devices to communicate and interoperate with one another. The goal is to publish standards, recommended practices, or guides that have broad market applicability and deal effectively with the issues of coexistence and interoperability with other wired and wireless networking solutions. As Section 3.4.2 will show, there are several study groups within the 802.15 group addressing different issues.

- *IEEE 802.11X WLAN:* The IEEE 802.11 standard [36] offers several WLAN technologies for use in the unlicensed 2.4- and 5-GHz bands. Legacy 802.11 systems operate in 2.4-GHz band with three different PHY layers sharing the same MAC layer. These PHY-layer specifications are the seldom-used infrared technology and the more popular DSSS and FHSS systems, achieving 1- and 2-Mbps data rates. Operating under the same 802.11 MAC layer in the 2.4-GHz band, higher data rates of 5.5 and 11 Mbps are supported by the IEEE 802.11b PHY-layer specification. The recent task group IEEE 802.11g has been formed to draft a standard that achieves data rates higher than 22 Mbps. Alternatively, in the 5-GHz band, the IEEE 802.11 standard offers the 802.11a specification that uses OFDM, achieving data rates of up to 54 Mbps. The 802.11e task group has been created to accommodate additional QoS provisions and security requirements at the MAC layer while supporting all of the previously mentioned legacy 802.11 PHY layers.

- *UWB:* UWB technology is loosely defined as any wireless transmission scheme that occupies a bandwidth of more than 25% of a center frequency or more than 1.5 GHz. Its emergence has mainly been associated with radar-based technologies. However, due to the recent development of high-speed switching technology, UWB is becoming increasingly attractive for low-cost consumer applications in future short-range wireless networks, that is, WLANs and PANs [37]. Three key features of the UWB favor this technology as a serious candidate for future PANs: great spatial capacity, potential compliance with global unlicensed operation, and implementation advantages.

At this point it is suitable to digress slightly to discuss the relationship between the WLAN and WPAN concepts. Namely, all short-range wireless technologies fall into two broad, overlapping categories [17]: PANs and LANs.

WPAN technologies emphasize low cost and low power consumption, usually at the expense of range and peak speed. WLAN technologies emphasize higher peak speed and longer range at the expense of cost and power consumption. Typically, WLANs provide wireless links from portable laptops to a wired LAN via APs, which is not a fully ad hoc feature. Regarding the isochronous traffic, the proposed IEEE 802.11e additions do not include guaranteed QoS in the ad hoc connection mode, which is paramount for high-rate WPAN applications.

Although each technology is optimized for its target applications, no hard boundary separates how devices can use WPAN and WLAN technologies. In particular, as Figure 3.3 shows, both could serve as a data or voice access medium to the Internet, with wireless WLAN technologies generally best suited for laptops, and WPAN technologies best suited for cell phones and other small portable electronics.

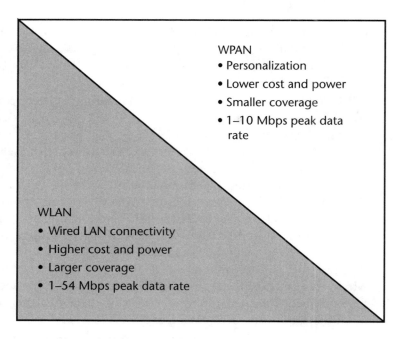

Figure 3.3 WLANs and WPANs' characteristics.

3.3.1 Comparison of Short-Range Wireless Technologies

Table 3.1 compares most important technical features for the existing technologies that are close to the WPAN paradigm: IrDA, HomeRF, Bluetooth, and IEEE 802.11. UWB is still not considered an actual specification for short-range technology and is thus not included. But, it has already been mentioned [37] that the potential of this technology is very high as Figure 3.4 illustrates.

3.4 Ongoing Research

This section presents the ongoing research in areas essential for the emergence of the WPAN paradigm. It tackles the architectural and middleware issues, problems related to data link design, ad hoc routing, and gateway functionality. Other important issues such as security are not observed.

3.4.1 Architecture and Middleware Issues

Wireless network deployment issues can be best understood if placed into the following three service classifications:

- WPAN;
- WLAN or wireless campus area networking;

Table 3.1 Overview of the Various Technologies Available Today Based on Their Usage, Cost, and Technological Merits

	IrDA	HomeRF	Bluetooth	IEEE 802.11		
Primary application	Widespread low-cost cable-replacement technology, mainly intended for data	Voice and data for home networking applications, remote control, and automation	Started as cable-replacement, but tends to gain more networking functionalities	Wireless Ethernet, AP for local data communication		
Frequency band	Optical infrared, wavelength 850 nm	2.4-GHz ISM	2.4-GHz ISM	2.4-GHz ISM and 5-GHz UNII		
Data rates	Typically 9.6–115 Kbps through serial ports; extensions offer 1-, 4-, and a maximum 16-Mbps data rates	0.8 and 1.6 Mbps, planned future iterations for up to 10 Mbps	Variable to maximum 732.2 Kbps	Standard	Maximum	Average
				Old .11	2 Mbps	1.2 Mbps
				.11b	11 Mbps	5.5 Mbps
				.11a	54 Mbps	24 Mbps
PHY	Optical	FHSS, 50 hops/sec	FHSS 1,600 hops/sec, 79 frequency channels at 1-MHz spacing	Standard	Type of PHY	
				Old .11	FHSS, DSSS, Ir	
				.11b	DSSS	
				.11a	OFDM at 5 GHz	
MAC	Polling	CSMA/CA for data, TDMA for voice	Master-slave, TDMA with polling	CSMA/CA in DCF and polling in PCF		
Range	Up to 2m, less for higher data rates; inhibited by obstacles, not omnidirectional	50m	10m (up to 100m with increased power)	50m indoors, 100m outdoors		
Topology	Master–slave	Peer-to-peer, MS-to-BS	Master-slave, real ad hoc network establishment without preconfiguration	Peer-to-peer in DCF, BS coordination in PCF		
Power management	By control of transmitting range	Connection point allows devices to enter power-saving mode	Defined low-power modes, possibility for flexible control of the low-power states (e.g., master–slave exchange)	AP schedules sleeping regimes of the nodes; in ad hoc DCF, low-power modes are not supported		
Security	Inherent physical security by limited range	56-bit Blowfish encryption	Differentiation of trusted and non-trusted devices, stream cipher, unique public address, two secret keys. Received critics about security weakness	WEP security protocol, but security improvement or additional security is needed		

Table 3.1 (continued)

Data network support	Via PPP	TCP/IP	Via PPP	TCP/IP
Voice support	RTCON is the specification for full-duplex voice support, but not simultaneously with data	32-Kbps ADPCM, voice network support via IP and PSTN, 1 Mbps still available under four voice connections	64-Kbps CVSD, maximum three simultaneous full-duplex voice connections in a piconet	Only in PCF mode are real-time services enabled
Coexistence	No interference	Interference with other technologies and transmitters in the ISM band	802.11 and 802.11b interfere with other ISM transmitters; 802.11a may be threatened if new PHY solutions move in UNII band	
Access to infrastructure LAN	Via IrLAN AP	SWAP-CA system is adapted from 802.11, gateway-like access to wired LAN	LAN access over PPP	By definition
Cost	Low	Medium	Medium to low	Medium to high
Advantages	Already highly present, lowest cost, lowest power consumption, no interference with existing LANs	Lower power requirement and cost than 802.11, higher speed than Bluetooth, conditional advantage of longer range	First "real" ad hoc technology, closest match to the PAN paradigm, low cost	High data rates, already wide deployment, fairly mature, conditional advantage of longer range
Disadvantages	Requires LOS, very short range for certain purposes	Unlikely to be established outside home environment; fails to become mainstream	Fairly low data rates, interference with other technologies in the ISM band, limited security	Significantly high cost, not scalable to personal devices, lacks backward compatibility

- WWAN.

Today, the core technology behind the wireless service in each of these classifications is unique and, more importantly, not an inherently integrated seamless networking strategy. For example, a user of a PDA, connecting to the Internet via a WAN service provider will not be able to connect directly to a local area wireless service. Simply stated, these are different services with different hardware requirements and fundamentally different service limitations.

The proliferation of mobile computing devices, including laptops, PDAs, and wearable computers, has created a demand for WPANs. WPANs allow proximal devices to share information and resources. The mobile nature of these devices places unique requirements on WPANs, such as low power consumption, frequent

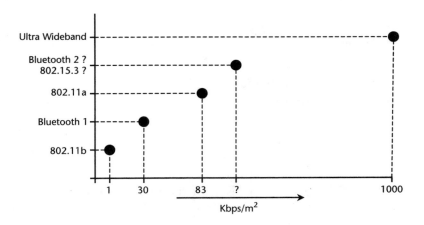

Figure 3.4   Comparison of IEEE 802.11, Bluetooth, and UWB.

make-and-break connections, resource discovery and utilization, and international regulations.

WPAN systems have evolved from "cord" replacement technologies. Some examples are as follows:

- Cordless communication between your keyboard and computer;
- Cordless communication between your personal productivity device or PDA and your computer;
- Cordless communication within your home between your cell phone and your home phone.

The most hyped of all WPAN wireless technologies today, Bluetooth, is a product of the telecommunications and computer industry Bluetooth Special Interest Group (SIG) and is rapidly gaining wide acceptance throughout the industry.

3.4.1.1   Current WPAN Architectural Models

Portable computing devices with wireless short-range links are seen as a new paradigm for computing and communication. The availability of low-cost, low-power, short-range international 2.4-GHz digital radios will provide the required communication technology for WPAN. Convergence of the competing radio standards would allow for a seamless connection among the digital wireless devices in our homes, offices, and environment. Spontaneous networks and service discovery and delivery are vital to the usefulness of PAN devices. A browser can point to any Web page on the Internet—a WPAN device should likewise be able to plug in to innumerable services anywhere in the world.

WPAN Versus WLAN Architecture
WPANs are different from WLANs in several functional areas:

- Control of the media;
- Range;
- Power;
- Number of participants;
- Ownership of devices;
- Nature of devices;
- Lifetime of network;
- Relative cost.

These differences are a result of the topological differences between the two types of these networks. From Figure 3.5 we can see that a WLAN is outwardly looking in the following ways:

- Interaction with wired infrastructure (LANs);
- Network time frame hours to days;
- Portable devices;
- "Wires are expensive."

A WPAN, however, is inwardly looking:

- Interaction with personal objects;
- Network time frame seconds to hours;
- Highly mobile devices;
- "Wires get in the way."

A WPAN Architectural Model Based on Bluetooth Technology
A Bluetooth configuration has the following characteristics and components (see Figure 3.6):

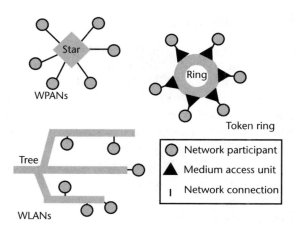

Figure 3.5   WPAN versus WLAN architecture.

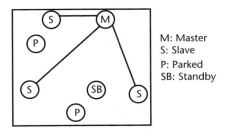

Figure 3.6 A Bluetooth piconet.

- It operates in the 2.4-GHz band at a data rate of 720 Kbps.
- It uses FHSS, which divides the frequency band into a number of channels (2.402–2.480 GHz, yielding 79 channels).
- Radio transceivers hop from one channel to another in a pseudorandom fashion determined by the master.
- It supports up to eight devices in a piconet (one master and seven slaves).
- Piconets can combine to form scatternets.

A piconet comprises the following:

- A number of devices are connected in an ad hoc fashion.
- One unit will act as a master and the others as slaves for the duration of the piconet connection.
- The master sets the clock and hopping pattern.
- Each piconet has a unique hopping pattern or ID.
- Each master can connect to 7 simultaneous or 200+ inactive (parked) slaves per piconet.

A scatternet has the following characteristics (see Figure 3.7):

- A scatternet is the linking of multiple colocated piconets through the sharing of common master or slave devices.
- A device can be both a master and a slave.
- Radios are symmetric (the same radio can be a master or slave).
- In a high-capacity system, each piconet has maximum capacity (720 Kbps).

In Figure 3.7 two interconnected piconets form a scatternet. They are able to operate within the vicinity of each other because they are using different hopping sequences, reducing mutual interference to an acceptable level. In this way it is possible to have several small groups of Bluetooth devices communicating with each other in the same area, which is particularly useful, for example, at a conference where individuals may be comparing notes while the main discussion points are being broadcast to all. Devices communicating using Bluetooth can transmit and

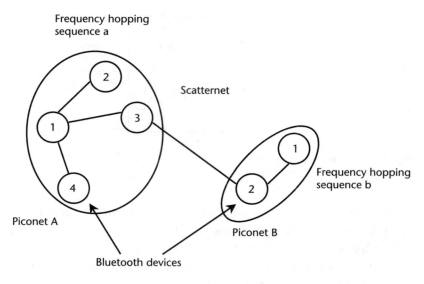

Figure 3.7   A Bluetooth scatternet.

receive at a rate of up to 1 Mbps, although in reality allowing multiple applications to communicate simultaneously will lower data rates. Bluetooth devices that are not currently part of a piconet are constantly "listening" for other Bluetooth devices, and when they are close enough to become part of a piconet, they identify themselves so that other devices can communicate with them if required. An example would be a Bluetooth-equipped printer and notebook PC. When the PC comes into range of the printer (arriving at the office, for example), the printer makes itself known to the PC so that when the user wishes to print a document, the two devices can immediately begin the data transfer. Meanwhile other PCs will have joined the piconet so that they too can use the printer when required.

A General WPAN Architectural Model

This section presents more general architectural approaches that are not specific to Bluetooth technology.

Figure 3.8 depicts an independent short-range radio system sample configuration. We can see that various kinds of devices can participate in such a configuration if they are supplied with the appropriate interfaces.

The above wireless links can be implemented with various current technologies; however, the dominant one is the Bluetooth standard. Today, the industry provides many hardware devices such as interface cards and other components for this technology (wireless Bluetooth PC card, wireless Bluetooth USB adapter).

This model accomplishes a peer-to-peer communication with the following characteristics:

- Cohabitation of multiple networks (20 or more);
- Up to 10 devices in a single WPAN;

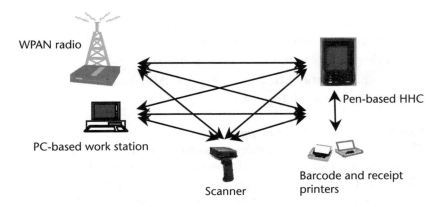

Figure 3.8 A WPAN architectural model.

- Dynamic WPAN and device IDs with network initiation;
- Network maintained devices coming and going;
- Support for temporary devices.

Although an independent network architecture may look ideal for an internal environment (e.g., the home), it is cut off from the rest of the world. So, in contrast a network infrastructure by which the WPAN devices can easily be connected to the outside world via an AP has been proposed.

In general an AP is a device that allows wireless-equipped devices to connect to a wired network. Figure 3.9 shows a limited infrastuctured network with the following characteristics:

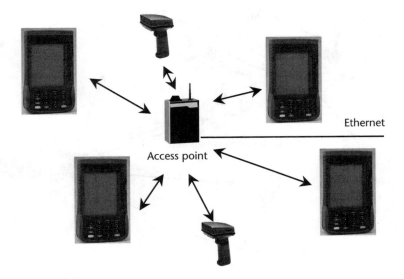

Figure 3.9   A limited infrastructured network.

- Main device (AP) having power at all times (fast access);
- Support for more than 10 devices;
- Ethernet APs with higher-layer protocols;
- Communications to the NT base (STAR base);
- Ability to switch from WPAN to LAN and back.

Support for Multimedia Terminals

The 3G terminals will provide access to many different forms of information and communication such as Web browsing, e-mail transmission and reception, video (slow scan for videophone-type connections and higher quality for short video clips and still pictures), and of course voice, making them true multimedia terminals. Voice will remain a major form of communication for humans, and this is recognized in the Bluetooth specifications by providing specific support for high-quality (64 Kbps) speech channels. With the ability to support packet data as well as speech (simultaneously, if required), Bluetooth can provide full local support for these multimedia applications. Bluetooth transceivers can support multiple data connections and up to three voice connections simultaneously, providing the functionality for a three-handset, cordless, multimedia/intercom system. Figure 3.10 shows three terminals in voice-intercom mode (one of which is a full multimedia terminal) and the alternative of two terminals in a conference speech connection with an outside connection. The limit of three interconnected terminals applies specifically to speech; the limit for the number of terminals exchanging data (Web connections, e-mail, and video depending upon bit-rate requirements) would be the upper limit of

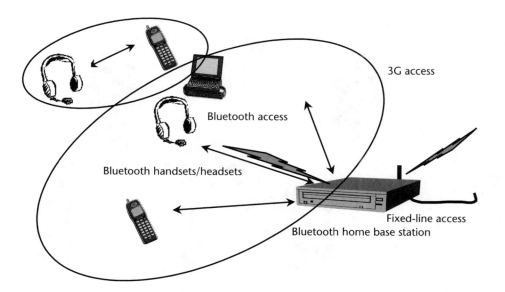

Figure 3.10   WPAN and the outside world.

eight per piconet. A Bluetooth-enabled home base station would provide the inter-connection for the local Bluetooth-equipped terminals and also for telephone "line" (may be a 3G terminal) connection. When an external connection is required, only two local handsets can participate in voice connections. The home base station forms a gateway between the home environment and national networks and services. Any two handsets can connect to each other, of course, without the home base station unit, making the whole arrangement very flexible.

This example highlights the complementary functionality of Bluetooth and 3G cellular systems. The 3G system is used to deliver a "trunk" connection to a specific location, and Bluetooth is used as a final delivery and local network connection. This will considerably reduce unnecessary traffic on the 3G network, creating a cost-effective solution for the convergence of fixed and mobile services and, as a by-product, keeping RF interference to a minimum.

WPAN Architectural Models in Working Environments

Working environments, where millions of people spend most of their day, combine several wireless technologies in order for WPAN devices to have access to remote applications.

In the architecture shown in Figure 3.11, factory machinery or tooling can move around the office premises in order to support WPAN device communication with other equipment that follows different standards, providing also the opportunity to access 3G services via cellular networks. In this case the machinery is used not only as an AP, but also translates protocols from different networks. The cellular wire-less phone acts like a gateway for WAN access.

The models in Figure 3.11 and the need to introduce 3G applications to WPAN networks demonstrate how essential intercommunication is. This trend makes com-ponents like home base stations and gateways indispensable in order to accomplish an adequate network infrastructure.

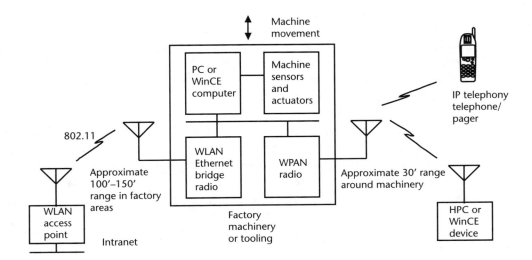

Figure 3.11   Factory machinery equipment.

WPANs and Gateways

An evolution is ongoing in terms of the residential gateway. Many in the design industry agree that this device will be an important piece of the puzzle, bringing ubiquitous high-speed connectivity to that great, untapped market, the home. But given the array of access options available to consumers—DSL, satellite, broadband wireless, even (gulp!) dial-up added to several data-distribution options within the home—and you've got a mélange of utter confusion.

Knowing that the residential gateway may prove to be the critical link in finally getting all of the amazing communication technology we know and love into a true mass market, we at Communication Systems Design discussed the design of this device with several prominent original equipment manufacturer (OEM) and component manufacturers and, as expected, heard a wide range of opinions as to what should be included.

Some of the concerns ring familiar: A gateway device should be equipped to accommodate changing standards. It may be designed to sit outside the home in a utilitarian sense or as a piece of furniture, such as we've seen with PCs in the past few years. Will the gateway take the form of a set-top box, or will it be something we haven't seen before? Part of the answer lies in visualizing what form tomorrows' home networks will assume.

Many wireless gateways have already been produced by several companies and are able to support many technologies, wired and wireless.

A good paradigm is Cisco's A85*xxx* series universal gateways, which can interface with all of the current transmission technologies. Bluetooth Network APs, such as WIDCOMM's BlueGate, provide secure gateways to the Internet, e-mail, and corporate LANs. BlueGate creates local hotspots of high-speed wireless connectivity that link as many as seven Bluetooth-enabled devices to any standard broadband modem (DSL, cable, or ISDN) through an Ethernet interface.

Thanks to such Bluetooth gateway devices, numerous network applications that no other technology can perform become possible. One example is a three-in-one phone. On the road, it's a conventional digital mobile phone. At home, it's a cordless phone with a connection to a home AP, linked to a fixed-wire phone network. At work, it's an extension of the desktop private branch exchange (PBX) phone. Additional examples of applications include automated ticket purchase using a phone or PDA in close proximity to a ticket-issuing machine, automated hotel check-in, and automated point-of-sale transactions.

Another commercial product comes from Wireless Networks. The BlueLAN is a LAN AP that allows multiple Bluetooth devices to access a local network through the BlueLAN's Ethernet port. This new technology allows Bluetooth-enabled mobile users to send and receive e-mails, surf the Web, and access other LAN and WAN resources. In addition, the company is designing DHCP, authentication, network address translation, and a PPP server into the LAN AP.

Wireless Networks deployed NetSilicon's complete open-source connectivity solution that runs uClinux, a form of the embedded Linux operating system, on the NetSilicon hardware platform.

Bluetooth and Gateways   Companies provide us today with many wireless gateways, which can be used in a WPAN topology. In order to provide a WPAN with an interface to a larger network (e.g., an 802.11 network), we should think of the following:

- Gateway devices to interconnect with Bluetooth devices;
- Gateway devices to interconnect with IEEE 802.11 (if the 802.11 standard not used as part of the WPAN standard).

Bluetooth provides a single point of entry at home, in the office, or abroad (see Figure 3.12). By eliminating the various wires needed to connect to other networks at the point of access, Bluetooth connects the user's appliance of choice wirelessly to the nearest Bluetooth network AP, allowing him or her to access the Internet through familiar procedures whether working at home, sitting by a pool, waiting for a flight in an airport lounge, or sipping latte in a neighborhood café.

Bluetooth turns the mobile phone into a communication gateway where information flows freely without boundaries (see Figure 3.13). Outside of the PAN, Bluetooth can be used as an instant AP facilitating the means for electronic commerce, personal finance, or data collection. Electronic payments at point-of-sale terminals, toll booths, or vending machines can be made by a Bluetooth mobile phone that instantly links the transaction to a personal bank account via a WAN for immediate processing. Similar financial transactions can also be performed at an interpersonal level by sending electronic currencies directly from one Bluetooth wallet to another. On a grander scale, electronic messages such as train or flight schedules and news flashes can be broadcast from special Bluetooth hot zones that are picked up by the public.

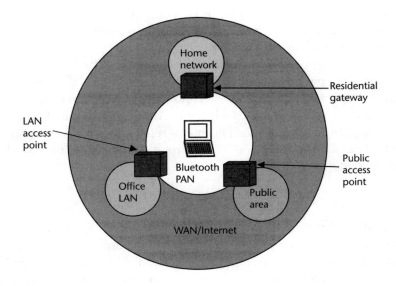

Figure 3.12   A general WPAN architectural model.

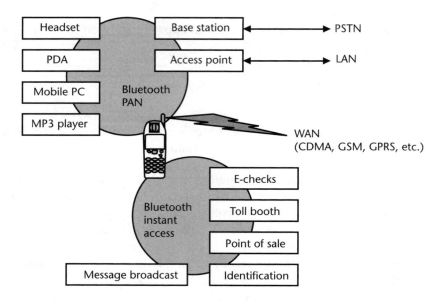

Figure 3.13   Bluetooth and APs.

Inspired by the above discussion, we can envision a set of services that a mobile entity (a user or a car) can access from a given spatial position. As the user walks around his premises, he gains access to different devices and services. We can think of those devices and the devices or sensors that the user is carrying as forming subsequent piconets.

The architectural model in Figure 3.14 can be envisaged as a zoom-in of the model depicted in Figure 3.12. If we combine the two figures, it becomes obvious that the case requires further investigation as it combines various technologies, components, and concepts.

A lot of effort has gone into the development of coexistence mechanisms to facilitate the parallel operation of WPAN and WLAN devices. Problems due to the parallel operation of these two technologies include interference as all of these systems use the unlicensed 2.4-GHz band. Although adopting frequency-hopping (FH) schemes can easily supercede problems related to interference, other issues come to the foreground in a general WPAN topology in which the user has the ability to access remote services.

3.4.1.2   Open Issues for Further Research

The issues that arise from general WPAN architectures involve the gateway component, which has to perform several functions, especially when the user is roaming inside his premises (interconnected piconets). Also the APs (home base stations) are components under discussion, and the communication between them remains an open issue. Furthermore, mobility scenarios and their management are very important when we consider scenarios like that presented in Figure 3.14.

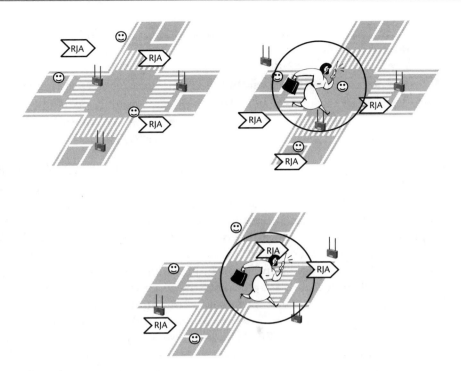

Figure 3.14   A mobile user and access to different services.

Bluetooth Mobility and Roaming

As the prevailing technology in WPANs so far is Bluetooth, this section presents recent research into user mobility. Two major current limitations of Bluetooth are that communication between devices must be direct (and hence is limited by the QoS of the radio channel between them) and that they do not support the movement of an active terminal from one network interface device to another. Essentially, in Bluetooth a device in piconet A cannot communicate with another Bluetooth device in piconet B, even if the device in piconet B was formerly a member of piconet A.

The sequel will briefly examine several theoretical ways in which these interpiconet issues could be handled, and developed. The Bluetooth SIG is in the process of addressing these issues in new profiles; however, the material presented here is neither based on nor influenced by the profile discussions. Thus, the models presented here may not relate to the profiles or to any actual roaming implementation.

3.5   Research Issues for Future WPAN Technology

The new emerging communication technologies will be centered on the user, adapting to the user's preferences and improving his living and working environments. Future WPAN is envisioned as a communications paradigm with a high level of personalization and ubiquitous access to information on demand and in an ad hoc manner (opportunity driven) through personal and public networking resources.

Future WPAN should provide the user with the following communications capabilities:

- Wireless connectivity within the POS (in-bubble networking);
- Access to sensors and actuators (mobile, wearable, or fixed);
- Access to the other wireless/wired networks (e.g., Internet via LAN APs, 3G network, organization's intranet) or other POSs.

Several applications scenarios are foreseen for the WPAN environment. They should be seamlessly available wherever the user is, supported by a variety of future services. In order to realize future WPAN dynamic and adaptive networking concepts, research activities will continue to concentrate on the topics that will provide optimized architecture and protocol solutions. The requirements of WPAN connections will develop synergistically with the pervasiveness of computing in general. In general, the achieved ubiquity of future short-range wireless technologies will depend on their degree of adaptation to the computing environment: fast service and resource discovery, as well as fast autoconfiguration and seamless interworking with communications entities from wireless and wired network infrastructures. Hence, the design of the WPAN should essentially depend on the wider-coverage legacy wireless systems with respect to coexistence and compatibility. Most importantly, the WPAN will also play the role of a complementary and integrating technology with the other technologies, being the "closest meter" to the user.

Some relevant future research areas include the following:

- *QoS/multimedia support:* The support of multimedia services will most likely be required within and throughout the ad hoc WPAN. However, QoS will gain different flavor in future short-range wireless networks due to the emergence of unique applications. For instance, there may be applications for which the timeliness and speed of ad hoc connection establishment is critical, while the actual rate that the application uses over the already established connection may be quite modest.

- *Trade-off QoS/power efficiency:* It has been stressed repeatedly that low power consumption is paramount for technology used by handheld battery-powered devices. Therefore, WPAN systems should be flexible in a way that will enable them to trade-off QoS for power saving. For instance, recall that the inherent stochastic communications quality in a wireless ad hoc network makes the provision of service guarantees difficult. Thus, a power-aware protocol should not perform transmission when the packet is highly likely to be received in error (which can be predicted to some extent), despite the fact that the QoS guarantee requires the transmission of that packet. Eventually, as soon as the communication conditions are stable, the WPAN should adapt and provide the user with better QoS, for example, as one defined for the access network.

The example above is another illustration of cross-layer optimization, something that should be a fundamental approach in designing power-aware ad hoc communication systems. In this particular case, the increased cross-layer interaction consists of controlling the activity of an upper layer (QoS provision entity) with information from a lower layer (bad channel conditions). Many challenging research issues are produced by the need to build richer interfaces among the layers of the protocol stack in wireless ad hoc systems. An optimal interplay of QoS and power efficiency can be only achieved if the optimization regards the protocol layers jointly.

- *Integration and cooperation with other networks:* Future WPAN should provide seamless integration and mobility management for heterogeneous infrastructures and other ad hoc networks. It will be IP oriented and should support the coexistence of IPv4 and IPv6. Mobility of terminal devices, within or with their POSs, addresses issues such as vertical and horizontal handover, location awareness, and roaming. The network infrastructures will strongly influence addressing, routing, and security solutions.

- *Interference/coexistence in unlicensed band:* At PHY and baseband conditioning, research efforts and solutions will largely depend on what is "low cost and simple" on the DSP market, thus determining the level of complexity and expense that will be acceptable for PAN transceiver solutions. Research into new radio interfaces has already started in Japan and Europe to increase the data rates of 3G mobile radio systems by more than one order of magnitude. Japan is proposing to reach at least 10 to 20 Mbps in a cellular environment and 2 Mbps in moving vehicles, using the multicarrier CDMA (MC-CDMA) as one of the candidates for multiple-access techniques. Generalized multicarrier CDMA (GMC) systems are capable of multiuser interference (MUI) elimination and ISI suppression, irrespective of the wireless frequency-selective channel encountered. Other license-free bands should be considered, such as the UNII 5 GHz or the unlicensed portion around 60 GHz. Finally, UWB is a highly promising research area because its transmission characteristics may be uniquely exploited by the upper layers (again cross-layer optimization).

- *Advanced link adaptation and MAC techniques:* The great volume of research into ad hoc routing protocols should be put in the context of concrete MAC-layer realization. Future work on short-range wireless networks must apply the routing protocols in a way that is adapted to both channel access/transmission conditions and application requirements.

The research into MAC-layer development for PAN should focus on cross-layer optimization, ensuring that the MAC-layer mechanisms include functionality to ensure guaranteed service levels (QoS) and power efficiency. This requires awareness of the MAC layer of the requirement put on the data streams originating from the network layer, as well as mechanisms for adjusting the parameters of the physical link. This means that the MAC layer should be able to exchange information with both the network layer and the PHY.

The requirement to offer QoS also implies the implementation of a MAC layer that is able to support different service classes, each with their own characteristics. Typically, some kind of scheduling is required for this; however, this is in contrast to the random nature of most current MAC protocols.

Actually, the scheduled users could override the best-effort users, but a delicate trade-off is required in this case. There are still many open issues when it comes to the development of random-access protocols that differentiate according to user class.

Especially in the case of multihop connections, the assurance of end-to-end QoS levels becomes a delicate issue. The mobility of the users makes a reservation scheme difficult to maintain; however, in the case of a WPAN of limited scale, a specialized solution might be developed.

- *Context discovery protocols and initialization techniques:* The area of context discovery and initialization will concentrate on developing different initialization algorithms and discovery protocols, such as location discovery, single-device discovery, discovery of infrastructure and noninfrastructure networks, service discovery, and environment discovery (e.g., security aspects).

- *Protocols and techniques for self-organization:* The areas of self-organization and reconfigurability for the support of wireless and mobile communications has recently received increased attention. The protocols and algorithms for self-organization should be combined with energy-aware routing and cooperative information-processing techniques.

- *Robust and optimized protocol stack:* The overall protocol stack should be optimized toward power consumption, rate-adaptation capability, end-to-end QoS, improved TCP/IP performances, and security requirements. This includes development of cross-layer optimization techniques, distributed TCP-aware PEPs, new intelligent PEPs implementation, development of a robust and adaptive protocol that can cope with packet losses, and bandwidth adaptation.

- *Security:* With the introduction of many wireless systems and the users of those different systems, the security of those connections and of the network, as well as authentication, become extremely important. The exact implementation of the security mechanism influences the design of the network architecture and should also be taken into account when developing Link layer and Network layer protocols. Complete stand-alone operation is difficult to achieve because all security protocols rely on a shared secret, which is usually exchanged in a different manner. For example, many security mechanisms known today rely on common access to a third party; however, in WPAN networks such a third party is not necessarily available.

The overall concept of WPAN in the future will be developed and enriched to include unthinkable solutions for new applications, services, and higher data-bit rates.

References

[1]   Pereira, J. M., "Fourth Generation: Now, It Is Personal," *Proc. PIMRC 2000*, September 2000, pp. 1009–1016.

[2]   Bhagwat, P., et al., "System Design Issues for Low-Power, Low-Cost Short-Range Wireless Networking," *IEEE Intl. Conf. Personal Wireless Communications (ICPWC)*, 1999, pp. 264–268.

[3]   Toh, C. K., *Wireless ATM and Ad-Hoc Networks—Protocols and Architectures*, New York: Kluwer Academic Publishers, 1997.

[4]   Perkins, C., *Ad Hoc Networking*, Reading, MA: Addison-Wesley, 2001.

[5]   IETF MANET, at http://www.ietf.org/html.charters/manet-charter.html.

[6]   Raman, B., P. Bhagwat, and S. Seshan, "Arguments for Cross-Layer Optimizations in Bluetooth Scatternets," *Proc. Symp. on Applications and the Internet*, 2001, pp. 176–184.

[7]   Chen, B., et al., "Span: An Energy-Efficient Coordination Algorithm for Topology Maintenance in Ad Hoc Wireless Networks," *Proc. ACMMOBICOM Conf.*, Rome, Italy, July 2001.

[8]   Toh, C. K., "Maximum Battery Life Routing to Support Ubiquitous Mobile Computing in Wireless Ad Hoc Networks," *IEEE Communications Magazine*, Vol. 39, No. 6, June 2001, pp. 138–147.

[9]   Royer, E. M., and C. K. Toh, "A Review of Current Routing Protocols for Ad-Hoc Mobile Wireless Networks," *IEEE Personal Communications*, Vol. 6, No. 2, April 1999, pp. 46–55.

[10] Feeney, L. M., *A Taxonomy for Routing Protocols in Mobile Ad Hoc Networks*, SICS Technical Report T99/07, October 1999.

[11] Stajano, F., and R. Anderson, "The Resurrecting Duckling: Security Issues for Ad Hoc Wireless Networks," in *Security Protocols, 7th International Workshop Proceedings, Lecture Notes in Computer Science*, B. Christianson, B. Crispo, and M. Roe, (eds.), 1999.

[12] Zhou, L., and Z. J. Haas, "Securing Ad Hoc Networks," *IEEE Network Magazine*, Vol. 13, No. 6, November/December 1999, pp. 24–30.

[13] Frodigh, M., P. Johansson, and P. Larsson, "Wireless Ad Hoc Networking—The Art of Networking Without a Network," *Ericsson Review*, No. 4, 2000, pp. 248–263.

[14] Jini project, at http://java.sun.com/products/jini/vision.

[15] Zimmerman, T. G., "Wireless Networked Digital Devices: A New Paradigm for Computing and Communication," *IBM Systems Journal*, Vol. 38, No. 4, 1999, pp. 566–574.

[16] Richardson, P., "Personal to Global: Wireless Technologies, 2005–2010," *Research Brief*, February 23, 2001, Gartner.com, at http://www.gartner.com/DisplayDocument?id=325116.

[17] Leeper, D. G., "A Long-Term View of Short-Range Wireless," *IEEE Computer*, Vol. 34, No. 6, June 2001, pp. 39–44.

[18] Rabaey, J., "PicoRadio Networks: An Overview," *Berkeley Wireless Research Center Focus 2000 Session*, July 2000.

[19] Zimmerman, T. G., "Personal Area Networks (PAN): Near-Field Intra-Body Communication," Master's thesis, MIT Media Lab, June 1995.

[20] Weiser, M., and J. Seely Brown, "Designing Calm Technology," *PowerGrid Journal*, Vol. 1.01, July 1996, at http://powergrid.electriciti.com.

[21] Daoud, F., and S. Mohan, "Service Portability and Virtual Home Environments," *IEEE Communications*, Vol. 40, No. 1, January 2002, pp. 76–77.

[22] Gupta, S. K., et al., "An Overview of Pervasive Computing," *IEEE Personal Communications*, Vol. 8, No. 4, August 2001, pp. 8–9.

[23] Wireless World Research Forum, at http://www.wireless-world-research.org.

[24] BlueDrekar Wireless Project, at http://www.research.ibm.com/BlueDrekar.

[25] Pervasive Computing at IBM India Research Lab, at http://www.research.ibm.com/irl/projects/pervasive.

[26]   Wireless Communication Systems at IBM Zurich Research Lab, at http://www.zurich. ibm.com/cs/wireless/index.html.

[27]   Advanced Network Technologies Division, at http://w3.antd.nist.gov.

[28]   WINLAB Focus Projects, at http://www.winlab.rutgers.edu/pub/docs/focus.

[29]   Terminodes Project, at http://www.terminodes.org.

[30]   MultiHop Project, at http://www.comnets.rwth-aachen.de/~ftp-mul.

[31]   WACNet Project, at http://www.acr.atr.co.jp/acr/general/dept1/TOP1/WACNet/WACNet frame1-E.htm.

[32]   Multihop Wireless Network, at http://www.ise.chuo-u.ac.jp/TISE/research_new/big/11shi-noda.html (Japanese only).

[33]   Autonomously Distributed Wireless Multihop Network (Kyoto University), at http://www. lab14.kuee.kyoto-u.ac.jp/adhoc_intro.htm (Japanese only).

[34]   Infrared Data Association (IrDA), at http://www.irda.org.

[35]   HomeRF Working Group, at http://www.homerf.org.

[36]   IEEE 802.11 Local and Metropolitan Area Networks: Wireless LAN Medium Access Control (MAC) and Physical Specifications, ISO/IEC 8802-11:1999, at http://grouper.ieee. org/groups/802/11.

[37]   Forester, J., et al., "Ultra-Wideband Technology for Short- or Medium-Range Wireless Communications," *Intel Technology Journal,* 2nd quarter, 2001, http://developer.intel. com/technology/itj/q22001/articles/art_4.htm.

[38]   Karagouz, J., "High-Rate Wireless Personal Area Networks," *IEEE Communications Magazine,* Vol. 39, No. 12, December 2001, pp. 96–102.

[39]   Williams, S., "IrDA: Past, Present and Future," *IEEE Personal Communications,* Vol. 7, No. 1, February 2000, pp. 11–19.

[40]   Prasad, A. R., "WLANs: Protocols, Security, and Deployment," Ph D. thesis, Delft University Press (DUP), Delft, the Netherlands, 2003.

[41]   Lansford, J., and P. Bahl, "The Design and Implementation of HomeRF: A Radio-Frequency Wireless Networking Standard for the Connected Home," *Proc. of IEEE,* Vol. 88, No. 10, October 2000, pp. 1662–1676.

[42]   Negus, K. J., A. P. Stephens, and J. Lansford, "HomeRF: Wireless Networking for the Connected Home," *IEEE Personal Communications,* Vol. 7, No. 1, February 2000, pp. 20–27.

[43]   Bluetooth SIG, at http://www.bluetooth.com.

[44]   Haartsen, J. C., "The Bluetooth Radio System," *IEEE Personal Communications,* Vol. 7, No. 1, February 2000, pp. 28–36.

[45]   Miller, B. A., and C. Bisdikian, *Bluetooth Revealed: The Insider's Guide to an Open Specification for Global Wireless Communications,* Upper Saddle River, NJ: Prentice Hall, 2001.

[46]   Bhagwat, P., "Bluetooth: Technology for Short-Range Wireless Apps," *IEEE Internet Computing,* Vol. 5, No. 3, May–June 2001, pp. 96–103.

[47]   Bluetooth SIG, Personal Area Networking Profile, revision 0.95a, at http://www.blue-tooth.com/dev/specifications.asp.

[48]   Johansson, P., et al., "Rendezvous Scheduling in Bluetooth Scatternets," *IEEE Int. Conf. Comm. (ICC),* Vol. 1, 2002, pp. 318–324.

[49]   Johansson, P., et al., "Bluetooth: An Enabler for Personal Area Networking," *IEEE Network Magazine,* Vol. 15, No. 5, September/October 2001, pp. 28–36.

[50]   Salonidis, T., et al., "Distributed Topology Construction of Bluetooth Personal Area Networks," *Proc. of IEEE INFOCOM 2001,* Vol. 3, April 2001, pp. 1577–1586.

[51]   Law, C., A. K. Mehta, and K. Y. Siu, "A New Bluetooth Scatternet Formation Protocol," *International Journal on Mobile Networks and Applications,* Vol. 8, No. 5, October 2003, pp. 485–498.

[52]   Ultra-Wideband Working Group, at http://www.uwb.org.

[53]   Rofheart, M., "XtremeSpectrum Multimedia WPAN PHY," IEEE 802.15.3 Working Group Submission, La Jolla, CA, July 2000, at http://grouper.ieee.org/groups/802/15/pub/Download.html.

[54]   Cover, T. M., and J. A. Thomas, *Elements of Information Theory*, New York: John Wiley & Sons, 1991.

[55]   FCC Notice of Proposed Rule Making, Revision of Part 15 of the Commission's Rules Regarding Ultra-Wideband Transmission Systems, ET-Docket 98-153.

[56]   FCC First Report and Order, FCC 02-48, at http://hraunfoss.fcc.gov/edocs_public/attach-match/FCC-02-48A1.pdf, May 30, 2002.

[57]   Siep, T. M., et al., "Paving the Way for Personal Area Network Standards: An Overview of the IEEE P802.15 Working Group for Wireless Personal Area Networks," *IEEE Personal Comm.*, February 2000, pp. 37–43.

[58]   Bluetooth SIG, Specifications of the Bluetooth System 1, ver. 1.0, December 1, 1999.

[59]   Karaoguz, J., "High-Rate Wireless Personal Area Networks," *IEEE Comm. Magazine*, December 2001, pp. 96–102.

[60]   Corson, M. S., J. P. Macker, and G. H. Cirincione, "Internet-Based Mobile Ad-Hoc Networking," *IEEE Internet Computing*, Vol. 3, No. 4, July–August 1999, pp. 63–70.

[61]   Lansford, J., A. Stephens, and R. Nevo, "Wi-Fi (802.11b) and Bluetooth: Enabling Coexistence," *IEEE Network*, Vol. 15, No. 5, September–October 2001, pp. 20–27.

[62]   Golmie, N., "Bluetooth and 802.11 Interference: Simulation Model and System Results," IEEE 802.15-01/195RO, April 2001.

[63]   Howitt, I., "WLAN and WPAN Coexistence in UL Band," *IEEE Trans. Vehicular Technology*, Vol. 50, No. 4, July 2001, pp. 1114–1124.

[64]   Lansford, J., "MEHTA: A Method for Coexistence Between Co-Located 802.11b and Bluetooth Systems," IEEE 802.15-00/360r0, November 2000.

[65]   Chiasserini, C. F., and R. R. Rao, "A Comparison between Collaborative and Non-collaborative Coexistence Mechanisms for Interference Mitigation in ISM Bands," *VTC 2001 Spring, IEEE 53rd VTS*, Vol. 3, 2001, pp. 2187–2191.

[66]   Eliezer, Oren, "Non-Collaborative Mechanisms for the Enhancement of Coexistence Performance," IEEE 802.15-01/092, January 2001.

[67]   Shoemake, Matthew B., "Proposal for Power Control for Enhanced Coexistence," IEEE 802.15-01/081, January 2001.

[68]   Jain, S., and R. Mahajan, *Wireless LAN MAC*, Internal Report, May 2000.

[69]   Sobrinho, J. L., and A. S. Krishnakumar, "Quality-of-Service in Ad Hoc Carrier Sense Multiple Access Wireless Networks," *IEEE Journal on Selected Areas in Communications*, Vol. 17, No. 8, August 1999, pp. 1353–1368.

[70]   Perkins, C. E., *Ad Hoc Networking*, Reading, MA: Addison-Wesley, 2001.

[71]   Garcia-Luna-Aceves, J. J., and C. L. Fullmer, "Performance of Floor Acquisition Multiple Access in Ad-Hoc Networks," *ISCC '98*, June 30–July 2, 1988, pp. 63–68.

[72]   Muir, A., et al., "Group Allocation Multiple Access with Collision Detection," *Infocom '97*, Vol. 3, April 1997.

[73]   Schiller, J., *Mobile Communications*, Reading, MA: Addison-Wesley, 2000.

[74]   Tobagi, F. A., and L. Kleinrock, "Packet Switching in Radio Channels: Part II—The Hidden Terminal Problem in Carrier Sense Multiple Access Modes and the Busy-Tone Solution," *IEEE Trans. on Communications*, Vol. 23, No. 12, 1975, pp. 1417–1433.

[75]   Tobagi, F. A., and L. Kleinrock, "Packet Switching in Radio Channels: Part III—Polling and (Dynamic) Split-Channel Reservation Multiple Access," *IEEE Trans. on Communications*, Vol. 24, No. 8, 1976, pp. 832–845.

[76] Sidhu, G. S., R. F. Andrews, and A. B. Oppenheimer, *Inside AppleTalk,* 2nd ed., Reading, MA: Addison-Wesley, 1980.

[77] Bertsekas, D., and R. Gallager, *Data Networks,* 2nd ed., Upper Saddle River, NJ: Prentice Hall, 1992.

[78] Ajmone-Marsan, M., and D. Roffinella, "Multichannel Local Area Networks Protocols," *IEEE Journal on Selected Areas in Communications,* Vol. 1, 1983, pp. 885–897.

[79] Lin, C. R., and J. S. Liu, QoS Routing in Ad Hoc Wireless Networks, *IEEE Journal on Selected Areas in Communications,* Vol. 17, No. 8, August 1999, pp. 1426–1438.

[80] Tseng, Y. C., et al., "A Multi-Channel MAC Protocol with Power Control for Multi-Hop Mobile Ad Hoc Networks," *21st Intl. Conf. on Distributed Computing Systems,* April 16–19, 2001, pp. 419–424.

[81] Nasipuri, A., J. Zhuang, and S. R. Das, "A Multichannel CSMA MAC Protocol for Multi Hop Wireless Networks," *Proc. of WCNC'99,* September 1999.

[82] Tang, Z., and J. J. Garcia-Lunes-Aceves, "Hop-Reservation Multiple Access (HRMA) for Ad Hoc Networks," *Proc. of Infocom'99,* October 1999, pp. 194–201.

[83] Mahonen, P., et al., "Platform-Independent IP Transmission over Wireless Networks: The WINE Approach," *IEEE Personal Communications,* Vol. 8, No. 6, December 2001, pp. 32–40.

[84] Balakrishnan, H., et al., "Improving TCP/IP Performance over Wireless Networks," *Proc. 1st ACM Conference on Mobile Communications and Networking (Mobicom),* Berkeley, CA, November 1995.

[85] Border, J., et al., "Performance Enhancing Proxies Intended to Mitigate Link-Related Degradations," IETF RFC 3135, June 2001.

[86] Wireless LAN Medium Access Control (MAC) and Physical Layer (PHY) Specifications, *The Institute of Electrical and Electronics Engineers, IEEE 802.11,* 1997.

[87] Santamaria, A., and F. López-Hernández, *Wireless LAN Systems,* Norwood, MA: Artech House, 1994.

[88] Radio Equipment and Systems (RES); High Performance Radio Local Area Network (HIPERLAN) Type 1; Functional Specification, *European Telecommunication Standard, ETS 300 652,* European Telecommunications Standards Institute, 1996.

[89] Broadband Radio Access Networks (BRAN); High Performance Radio Local Area Network (HIPERLAN) Type 1; Functional Specification, *EN 300 652 v1.2.1,* European Telecommunications Standards Institute, 1996.

[90] Rabaey, J. M., et al., "PicoRadio Supports Ad Hoc Ultra-Low Power Wireless Networking," *IEEE Computer,* Vol. 33, No. 7, July 2000, pp. 42–48.

[91] Tanenbaum, A. S., *Computer Networks,* 3rd ed., Upper Saddle River, NJ: Prentice Hall, 1996.

[92] Abramson, N., "The ALOHA System—Another Alternative for Computer Communications," *AFIPS,* Vol. 36, 1970, pp. 295–298.

[93] Golmie, N., N. Chevrollier, and I. ElBakkouri, "Interference Aware Bluetooth Packet Scheduling," *IEEE Globecom,* 2001.

[94] Walrand, J., *Communication Networks: A First Course,* 2nd ed., New York: McGraw-Hill, 1998.

[95] Perkins, C. E., and P. Bhagwat, "Highly Dynamic Destination-Sequenced Distance-Vector Routing (DSDV) for Mobile Computers," *Proc. of SIGCOMM'94 Conference on Communications, Architectures, Protocols and Applications,* August 1994, pp. 234–244.

[96] Murthy, S., and J. J. Garcia-Luna-Aceves, "A Routing Protocol for Packet Radio Networks," *Proc. of the 1st International Conference on Mobile Computing and Networking (ACM Mobicom),* Berkeley, CA, November 13–15, 1995, pp. 86–95.

[97] Pei, G., M. Gerla, and T. W. Chen, "Fisheye State Routing in Mobile Ad Hoc Networks," Internet Draft, draft-ietf-manet-fsr-00.txt, November 2000.

[98] Pei, G., M. Gerla, and T. W. Chen, "Fisheye State Routing in Mobile Ad Hoc Networks," *Proc. of IEEE International Conference on Communications (ICC)*, New Orleans, LA, June 2000, pp. 70–74.

[99] Kleinrock, L., and K. Stevens, *Fisheye: A Lens-Like Computer Display Transformation*, Technical report, UCLA, Computer Science Department, 1971.

[100] Royer, E. M., "Routing in Ad Hoc Mobile Networks: On Demand and Hierarchical Strategies," Ph.D. dissertation, Electrical and Computer Engineering, University of California, Santa Barbara, December 2000.

[101] Johnson, D. B., and D. A. Maltz, "Dynamic Source Routing in Ad Hoc Wireless Networks," in *Mobile Computing*, T. Imielinski and H. Korth, (eds.), New York: Kluwer Academic Publishers, 1996, pp. 152–181.

[102] Johnson, D. B., "Routing in Ad Hoc Networks of Mobile Hosts," *Proc. of IEEE Workshop on Mobile Computing Systems and Applications*, December 8–9, 1994, pp. 158–163.

[103] Johnson, D. B., D. A. Maltz, and Y. C. Hu, "The Dynamic Source Routing Protocol for Mobile Ad Hoc Networks (DSR)," Internet Draft, draft-ietf-manet-dsr-09.txt, April 15, 2003.

[104] Park, V. D., and M. S. Corson, "A Highly Adaptive Distributed Routing Algorithm for Mobile Wireless Networks," *Proc. of IEEE Conference on Computer Communications (INFOCOM)*, Kobe, Japan, April 1997, pp. 1405–1413.

[105] Park, V. D., and M. S. Corson, "Temporally-Ordered Routing Algorithm (TORA) Version 1, Functional Specification," Internet Draft, draft-ietf-manet-tora-04.txt, July 2001.

[106] Park, V. D., and M. S. Corson, "A Performance Comparison of TORA and Ideal Link State Routing," *Proc. of IEEE Symposium on Computers and Communication 1998*, June 1998.

[107] Haas, Z. J., M. R. Pearlman, and P. Samaq, "The Zone Routing Protocol (ZRP) for Ad Hoc Networks," Internet Draft, draft-ietf-manet-zone-04.txt, July 2002.

[108] Royer, E. M., and C. E. Perkins, "Multicast Ad Hoc On Demand Distance Vector (MAODV) Routing Protocol," *Proc. of 5th ACM/IEEE International Conference on Mobile Computing and Networking (MOBICOM)*, Seattle, WA, August 1999, pp. 207–218.

[109] Cheng, E., "On-Demand Multicast Routing in Mobile Ad Hoc Networks," M. Eng. thesis, Department of Systems and Computer Engineering, Carleton University, Ottawa, January 2001.

[110] Garcia-Luna-Aceves, J. J., and E. L. Madruga, "The Core-Assisted Mesh Protocol," *IEEE Journal on Selected Areas in Communications*, Vol. 17, No. 8, August 1999, pp. 1380–1394.

[111] Lee, S. J., W. Su, and M. Gerla, "On-Demand Multicast Routing Protocol (ODMRP) for Ad Hoc Networks," Internet Draft, draft-ietf-manet-odmrp-02.txt, January 2000.

[112] Prasad, R., "Basic Concept of Personal Area Networks," *WWRF, First Meeting*, Munich, Germany, 2001.

[113] Niemegeers, I. G., R. Prasad, and C. Bryce, "Personal Area Networks," *WWRF, Second Meeting*, Helsinki, Finland, May 10–11, 2001.

[114] Prasad, R., "60 GHz Systems and Applications," *2nd Annual Workshop on 60 GHz WLAN Systems and Technologies*, Kungsbacka, Sweden, May 15–16, 2001.

[115] Prasad, R., and L. Gavrilovska, "Personal Area Networks," keynote speech, *Proc. EUROCON*, Bratislava, Slovakia, Vol. 1, July 2001, pp. 3–7.

[116] Prasad, R., and L. Gavrilovska, "Research Challenges for Wireless Personal Area Networks," keynote speech, *Proc. of 3rd International Conference on Information, Communications and Signal Processing (ICICS)*, Singapore, October 2001.

[117] Munoz, L., and R. Prasad, *WLANs and WPANs towards 4G Wireless*, Norwood, MA: Artech House, 2003.

[118] Prasad, R., and M. Ruggieri, *Technology Trends in Wireless Communication*, Norwood, MA: Artech House, 2003.

CHAPTER 4

Appropriate Channel Model for OFDM Systems

4.1 Introduction

The channel model is the bread and butter for the telecommunications engineer.

This is how I used to emphasize in my lectures the importance of the channel model for designing radio interfaces for wireless communications systems. And this was not just a phrase to keep the students' attention. The channel models really are the foundation on which mobile communications systems are built.

As the main topic of this book is the design of OFDM air interfaces, this chapter describes and discusses an appropriate channel model for such systems [1–44]. This model must address two general requirements:

1. Analytical treatment of OFDM-related problems;
2. Efficient computer simulation schemes.

According to the system's key specifications, it should fit into physical radio channels in the millimeter-wave frequency band for indoor (in-room) and short-range outdoor environments.

Radio propagation in a mobile radio channel is determined mainly by its multi-path nature. Multiple reflections and sometimes an LOS component of the transmitted signal arrive at the receiver via different propagation paths and, therefore, with different amplitudes and delay times. As an effect of this, the narrowband received power fluctuates dramatically when observed as a function of location (or time) and frequency. In the early days of mobile systems, the communications engineer was mainly interested in the time-variability of narrowband channels, which were thus studied extensively (see, e.g., [1]). By that time, transmission bandwidths were small; thus, flat-fading was a reasonable assumption. As the systems have evolved, demand for higher transmission rates has been increasing, making the channel's time dispersion (which is equivalent to its frequency selectivity) a major issue.

In OFDM, the channel's variability in the frequency domain (FD) has a similar role as the time variance in a (flat-fading) narrowband system. Usually, the channel can be assumed to be static during the transmission of at least one OFDM symbol. In indoor WLANs, the channel is even considered quasistationary during up to a whole data-packet or frame period.

The following section reviews the propagation mechanisms that have to be characterized by the channel model. The main properties of interest for OFDM system design are emphasized, and important channel parameters are defined.

The so-called frequency-domain channel model (FD model) is proposed and analyzed in Section 4.3. The model describes frequency-selective fading by the delay power spectrum (DPS) of the channel and the Fourier transform (FT) of the spaced-frequency correlation function [2]. This approach defines the time variability of the Doppler power spectrum [3], which is often referred to as Jakes's fading model [1, 4]. Expressions are given relating the DPS (being specified by just two to four parameters) to the most important physical channel parameters. Rayleigh and Rician fading channels are considered.

A direct implementation of the FD channel model in a computer-simulation scheme is proposed in Section 4.4. The outputs of this simulator are (complex-valued) frequency-selective channel transfer functions. The differences are emphasized between this approach and (conventional) time-domain (TD) simulators, which generate channel impulse responses.

Section 4.5 summarizes the basic results of a number of measurement campaigns. Some of them were performed at Delft University of Technology (DUT) in the Netherlands; others were found in the literature. We elaborate on the suitability of the proposed channel model for describing the radio channels investigated.

Conclusions and recommendations are given in Section 4.6.

4.2 Characterization of the Mobile Radio Channel

This section starts with a qualitative description of the main propagation mechanisms resulting from multipath wave interference. Section 4.2.2 reviews mathematical definitions that are useful for the characterization of wideband, frequency-selective, mobile radio channels. Important channel parameters are introduced and their physical interpretation is developed.

4.2.1 Components of a Multipath Channel Model

For the mathematical description of a multipath radio channel, it is convenient to distinguish three mechanisms:

1. Path loss;
2. Shadowing;
3. Multipath interference.

The former two are described by large-scale channel models, which essentially provide information about the average received power at a certain location. Path loss strictly describes the dependency of this average power on the distance between transmitter and receiver, while shadowing accounts for the fluctuations observed at a fixed distance, due to geometric features of the propagation environment. These fluctuations occur for instance because of the blocking of relevant propagation paths (e.g., the LOS component) as the mobile moves around.

Highly sophisticated large-scale models that typically employ geographical information system (GIS) databases in order to account for topographical features are incorporated in the cell-planning tools used by mobile system operators. In indoor environments, ray-tracing models are often used to predict the received signal strength at a given location.

Such models are not relevant, however, for the design of new transmission techniques. For this application, a description of the effects of multipath interference is required since the air interface has to cope with them. These effects are often referred to as small-scale fading.

Small-scale models are valid within (small) local areas, where the signal fluctuations due to shadowing and path loss can be neglected. The dimensions of such a local area are therefore limited to approximately 5 to 40λ, where λ is the wavelength of the RF carrier. (Due to the small wavelengths below 1 cm, this range may be even larger in the millimeter-wave band.)

The channel model investigated in this chapter is limited to the description of small-scale effects. A set of average parameters specifies the channel's behavior within a local area. These parameters are the normalized received power (NRP),[1] P_0, the Rician K-factor, K, and the rms delay spread (RDS), τ_{rms}. Note, however, that each realization obtained from the model has varying instantaneous parameters denoted $\left\{\hat{P}_0, \hat{K}, \hat{\tau}_{rms}\right\}$ since the model is a stochastic one. (To be specific, it is a Gaussian wide-sense stationary uncorrelated scattering (WSSUS) model, as shown in Section 4.3 and [5, 6]). The amount of variation of these parameters from the local area parameters depends in particular on the observed bandwidth. When the bandwidth is much greater than the coherence bandwidth, then the multipath is completely resolved and the channel parameters vary little because the individual multipath amplitudes do not change rapidly within a local area. However, if the system is narrowband, then multipath is not resolved, and the path amplitudes at each resolvable (delay) time bin (being spaced by the reciprocal of the bandwidth) vary due to multipath interference. This leads to the fluctuation of the instantaneous channel parameters within the local area (see [6, 7]).

The mathematical definitions of the channel parameters are given in the following section. The behavior indicated above will be revisited based on the equations presented.

The selection of these parameters is an attempt to specify the main characteristics of the frequency-selective channel with a minimum number of variables.

4.2.2   Definitions

4.2.2.1   Channel Impulse Response

It is most illustrative to start with the definition of channel impulse response (IR), which is the straightforward formulation of the sum of discrete multipath

---

1.   The (dimensionless) NRP is defined as the ratio of the received power P_{rx} and the transmitted power P_{tx}. Equivalently, the absolute received power P_{rx} [W] could be used for channel description.

components impinging at the receiver. In complex lowpass equivalent notation, the IR is written as

$$h(\tau) = \sum_i \beta_i e^{-j\theta_i} \delta(\tau - \tau_i) \qquad (4.1)$$

where $\{\beta_i\}$, $\{\theta_i\}$, and $\{\tau_i\}$ are the propagation paths' amplitudes, phases, and delays, respectively, and τ is the delay-time variable. Normally, the delay of the first (shortest) ray is defined as $\tau_0 = 0$ because the absolute delay times are not important; only the time dispersion is. Therefore, τ is called the excess delay time, and it follows that $\tau_i > 0$ for $i > 0$; that is, the channel IR is causal.

Note that in a real environment, the parameters $\{\beta_i\}$, $\{\theta_i\}$, and $\{\tau_i\}$ are time variant. For the sake of simplicity, this time dependency was omitted in (4.1). Within a local area, that is, for displacements in the order of a few wavelengths λ, the ray amplitudes $\{\beta_i\}$ and the delays $\{\tau_i\}$ can be considered relatively static, corresponding to the assumption of a negligible change in shadowing. The ray phases $\{\theta_i\}$, however, change unpredictably within the interval $[0, 2\pi)$ because they are related to the absolute path lengths.[2]

4.2.2.2   Channel Parameters

All channel parameters introduced here are defined from the (static) power delay profile (PDP), which is a function derived from the channel IR (4.1). The PDP specifies the ray power versus the delay-time structure of the IR, being

$$p(\tau) = \sum_i \beta_i^2 \delta(\tau - \tau_i) \qquad (4.2)$$

As the ray phases are dropped in this equation, the channel parameters must be (largely) constant within the local area, provided that the propagation paths are fully resolvable.

The first parameter is the (normalized) received power, being the sum of the ray powers

$$P_0 = \sum_i \beta_i^2 \qquad (4.3)$$

The Rician K-factor is the ratio of the dominant path's power to the power in the scattered paths, defined as

---

2.   It is an open issue whether the assumption of discrete paths is viable. Generally, each reflection will show some time dispersion and, therefore, a frequency-dependent magnitude. However, for a given observation bandwidth, such physical paths can normally be approximated by (a number of) discrete Dirac impulses.
   Smulders states, based on channel measurements over bandwidths of 2 GHz in the 60-GHz band, that millimeter waves have sufficiently small wavelengths to be modeled as rays following discrete paths (see [8], 432 f.).

$$K = \frac{\beta_{i,\max}^2}{P_0 - \beta_{i,\max}^2}$$

where

$$\beta_{i,\max} = \max_i\{\beta_i\} \qquad (4.4)$$

It will be seen that the K-factor specifies the depth of the fades within a local area, as the Rician probability density function (PDF) will be used to characterize the amplitude distribution of the channel response. Larger K-factors relate to shallower fades.

Note that in the presence of an LOS, the first ray is the dominant one, implying that $\beta_{i,max} = \beta_0$ at $\tau_0 = 0$.

Finally, the RDS is introduced, which is the second central moment of the (power-normalized) PDP, written as

$$\tau_{rms} = \sqrt{\overline{\tau^2} - \overline{\tau}^2}$$

where

$$\overline{\tau^m} = \sum_i \tau_i^m \beta_i^2 / P_0, m = \{1, 2\} \qquad (4.5)$$

τ_{rms} is considered to be the most important single parameter for specifying the time extent of the dispersive channel. It also characterizes the frequency selectivity because τ_{rms} is related to the average number of fades per bandwidth and to the average bandwidth of the fades.

4.2.2.3 Channel Transfer Function

An equivalent description of the time-dispersive channel is obtained by applying the FT to the IR, yielding the channel transfer function (TF). This step will demonstrate that a time-dispersive channel is also frequency selective.

First, time variability is reintroduced to the IR (4.1),

$$h(\tau,t) = \sum_i \beta_i(t)e^{-j\theta_i(t)}\delta(\tau - \tau_i(t)) \qquad (4.6)$$

leading to the time-variant TF,

$$H(f,t) = \int_{-\infty}^{\infty} h(\tau,t)e^{-j2\pi ft}\,d\tau = \sum_i \beta_i(t)e^{-j[2\pi f\tau_i(t)+\theta_i(t)]} \qquad (4.7)$$

The magnitude of this function shows rapid variations with respect to both the time and frequency variables. $H(f,t)$ can be seen as the vector sum of the ray amplitudes $\{\beta_i(t)\}$, with vector angles $\{2\pi f\tau_i(t)+\theta_i(t)\}$. As the ray phases $\{\theta_i(t)\}$ change rapidly for small displacements, the vector sum changes, causing the location and time variability. The frequency dependency is due to the different delay times $\{\tau_i(t)\}$, which, at different frequencies, also lead to drastic changes in the vector sum. The phases at two specific frequencies differ more with larger excess delay times $\{\tau_i(t)\}$. This suggests a dependency of the time extent of the IR (which is characterized by the RDS) and the number of fades per unit of bandwidth.

4.2.2.4   Magnitude Distribution

Due to the quasirandom phases of the terms of (4.7), $H(f,t)$ can be seen as the sum of a (large) number of random variables (RVs) with amplitudes $\{\beta_i(t)\}$ and uniformly distributed phases over $[0, 2\pi)$. Assuming that a considerable number of rays have similar magnitudes (except for possibly one dominant ray), the central limit theorem leads to the conclusion that $H(f,t)$ has a complex Gaussian distribution. Without the dominant ray it is zero mean; otherwise it is nonzero mean. The magnitude $R = |H(f,t)|$ of the complex Gaussian process is described by the Rician PDF

$$p_R(r) = \frac{r}{\psi_0} e^{-\frac{(r^2+\rho^2)}{2\psi_0}} I_0\left(\frac{r\rho}{\psi_0}\right) \tag{4.8}$$

where ψ_0 is the common variance of the real and imaginary components of the complex Gaussian process, ρ is the amplitude of the mean of $H(f,t)$, $\rho = |E\{H(f,t)\}|$, and $I_0(\bullet)$ is the zeroth-order modified Bessel function of the first kind. For the zero-mean case ($\rho = 0$), the Rician PDF reduces to the Rayleigh PDF.[3]

The parameters of (4.8) are related to the channel parameters P_0 and K as

$$\rho^2 = \beta^2_{i,max} = P_0 \frac{K}{K+1} \text{ and } 2\psi_0 = P_0 - \beta^2_{i,max} = P_0 \frac{1}{K+1} \tag{4.9}$$

Note that ρ^2 is the power of the dominant component, while $2\psi_0$ is the power of the scattered components. If the central limit theorem (plus dominant path) is not perfectly valid, then the parameters given in (4.9) may still express a best fit of the Rician distribution to the given channel. However, in this case, the magnitude of the dominant path $\beta_{i,max}$ may rather be seen as an "equivalent" dominant path gain, which does not strictly relate to one physical propagation path.

4.2.2.5   Band-Limiting the TF and Sampling the IR

For computer simulation schemes, a sampled version of the channel IR is required, which implies the band limitation of the respective TF. Let us first introduce the

---

3. If the dominant component $\tau_{i,max}$ occurs at a delay time different from $\tau_0 = 0$ (or at a nonzero Doppler frequency), then the mean will become zero as well, as a (deterministic) complex harmonic component results. However, the amplitude distribution is still appropriately described by the Rician distribution. (This case is described by Rice as the "Distribution of Noise Plus Sine Wave" [9, 10].)

latter. A sampling interval T_s in the TD limits the bandwidth to $\pm\dfrac{BW}{2} = \pm\dfrac{1}{2T_s}$.

Multiplication of the TF (4.7) by a rectangular window $W_{BW}(f)$ applies such band limitation

$$H_{BW}(f,t) = H(f,t) \cdot W_{BW}(f)$$

where

$$W_{BW}(f) = \begin{cases} 1 & \text{if } |f| \le \dfrac{BW}{2} \\[2mm] 0 & \text{if } |f| > \dfrac{BW}{2} \end{cases} \qquad (4.10)$$

This step is equivalent to a convolution of the IR by a sinc function

$$h_{BW}(\tau,t) = h(\tau,t) * \operatorname{sinc}(\tau/T_s) = \sum_i \beta_i(t) e^{-j\theta_i(t)} \operatorname{sinc}\frac{\tau - \tau_i(t)}{T_S} \qquad (4.11)$$

where $\operatorname{sinc} x = \begin{cases} \dfrac{\sin \pi x}{\pi x} & \text{if } x \ne 0 \\[2mm] 1 & \text{if } x = 0 \end{cases}$. Clearly, rays stop being resolvable if the delay-time separation between adjacent rays is in the range of T_s or below.

Sampling in the TD can be seen as multiplication by a train of Dirac impulses with period T_s. It therefore has the effect of convoluting the FD representation by a pulse train with period $1/T_s = BW$ [11]. The prior band limitation keeps the thereby duplicated spectra from overlapping (i.e., aliasing is avoided), which essentially means that no information is lost through the sampling. The sampled IR becomes

$$h_{BW,n_\tau}(t) \sum_i \beta_i(t) e^{-j\theta_i(t)} \operatorname{sinc}\frac{n_\tau T_s - \tau_i(t)}{T_S} \qquad (4.12)$$

with $n_\tau = \{\ldots -1, 0, 1, 2, \ldots\}$ being the discrete delay-time index. From this equation, one can observe that the IR has contributions of all propagation paths at any time bin n_τ. (Except if a ray has an excess delay of $\tau_i = kT_s$, where k is an integer, to be exact). Even at negative delay times, some "leakage" of the (causal) IR is evident. From (4.12) it also becomes clear that, for limited time resolution or bandwidth, the sampled IR (at any time bin n_τ) is rapidly time variant, due to the time dependency of the superimposed rays' phases $\{\theta_i(t)\}$. Calculating channel parameters from this sampled IR results in instantaneous parameters $\{\hat{P}_0, \hat{K}, \hat{\tau}_{rms}\}$, which are time variant, even within a local area, as discussed in Section 4.2.1. The variability of these parameters is shown in Section 4.2.3, based on simulation results.

The application of the central limit theorem again leads to the conclusion that complex Gaussian processes appropriately model the coefficients $\{h_{BW,n_\tau}(t)\}$ (see

[6, 12]). Their variances follow the so-called average PDP, which usually decays with increased delay time. In various channel models, the IR is described in this way (see, e.g., [13–16]). The complex Gaussian distribution also applies to the ray gains of IRs derived from the FD channel model, which is proposed in Section 4.3.

The above analysis is an attempt to describe theoretically the behavior of the time-variant, frequency-selective radio channel. It focuses on the aspects that are important for a deeper understanding of the FD channel model. Therefore, particularly the frequency selectivity of a band-limited, quasistatic channel has been discussed.

4.2.3 Variation of Channel Parameters Due to Bandwidth Limitation

This section presents simulation results of the variability of instantaneous channel parameters within a local area.

Channel realizations were generated with a TD simulation scheme, which produces channel IRs. The simulation model assumes an LOS ray at $\tau = 0$, a Poisson process of ray arrivals (approximately 60 rays), an exponentially decaying average PDP, and complex Gaussian ray amplitudes (see [16] for one cluster; see Section 4.5.3). In a second step, the IRs were normalized to get the required K-factor K, $\tau_{rms} = 1$ and $P_0 = 1$ [15], allowing for simple evaluation of the estimation error. Applying the FT to the generated IRs, (complex-valued, discrete-frequency) TFs were obtained with arbitrary bandwidth.

The variability of $\hat{\tau}_{rms}$ and $\hat{P}_0$ within a local area is depicted in Figure 4.1. For analyzing $\hat{\tau}_{rms}$, the simulated TFs were transformed back to the delay TD using the inverse DFT (IDFT) without windowing. Consecutively, $\hat{\tau}_{rms}$ was determined from the positive part of the obtained, sampled channel IR using (4.5). $\hat{P}_0$ is simply the average power of the band-limited TFs.

The standard deviations of these parameters decrease with increasing bandwidth because individual "propagation" paths become gradually more resolvable. The estimation bias in $\hat{\tau}_{rms}$ [see Figure 4.1(a)] is due to leakage effects.

Reduced variance and bias for higher K-factor are intuitively explained by the fact that the (deterministic) dominant path largely determines Rician channels. Note that K has most influence on the instantaneous values of the average power P_0. This behavior can be anticipated because K directly relates to the depth of the fades; that is, a channel with a high K-factor (which has shallow fades) shows less variation in this parameter than, for instance, a Rayleigh fading channel (which has quite deep fades).

4.3 FD Channel Modeling

The channel model proposed in this section describes the correlation properties of the channel TF in the FD. Starting from the definition of the channel correlation functions (and power spectra), the so-called FD channel model is derived. Mathematical expressions are given, relating the model's parameters to (physical) channel parameters.

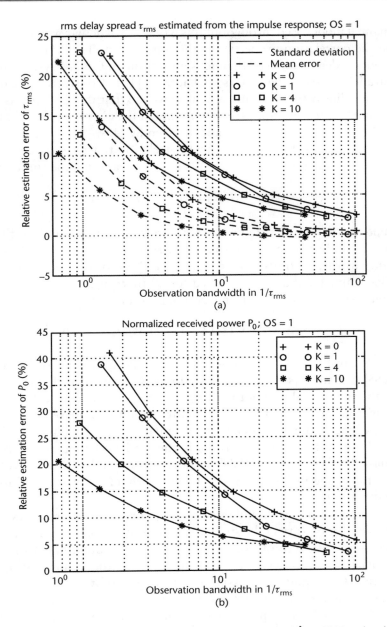

Figure 4.1   (a) Bias and standard deviation of the instantaneous RDS $\hat{\tau}_{rms}$ within a local area due to band limitation (the bias is caused by leakage effects); (b) standard deviation of the instantaneous, normalized, received power $\hat{P}_0$ within a local area.

4.3.1   The WSSUS Channel Model

The channel correlation functions and power spectra are a set of functions defining the small-scale characteristics of multipath fading channels in more detail than the channel parameters given above. Introducing some assumptions will lead to the channel model used throughout this work. In particular, we concentrate in this work on the correlation properties of the time-variant TF $H(f,t)$ [see (4.7)] because this function determines the channel's impact on an OFDM system modeled as a set of parallel Gaussian channels (see Chapter 5, [17]). Considering the mobile radio channel as a linear time-variant system, it is seen that the TF $H(f,t)$ is only one possible channel representation (from the family of Bello's system functions [5, 6]). Another is, for instance, the time-variant IR given by (4.6).

Let us first define the channel correlation functions assuming that those functions are wide-sense stationary (WSS). This means that the autocorrelation function

$$\phi_H(f_1,f_2,t_1,t_2) = E\{H^*(f_1,t_1)H(f_2,t_2)\} \tag{4.13}$$

depends only on the frequency separation $\Delta f = f_1 - f_2$ and on the time separation $\Delta t = t_1 - t_2$, but not on the absolute observation frequencies $\{f_1, f_2\}$ and times $\{t_1, t_2\}$. In other words, the time-variant TF $H(f,t)$ is WSS with respect to both variables f and t. The channel is thus characterized for all times and all frequencies by the so-called spaced-frequency, spaced-time correlation function

$$\phi_H(\Delta f, \Delta t) = E\{H^*(f,t)H(f + \Delta f, t + \Delta t)\} \tag{4.14}$$

It can be shown that this assumption is equivalent to the introduction of the WSSUS channel (see, e.g., [2, 5, 6]). In the WSSUS channel, the WSS property applies to the time variability of the IR $h(\tau,t)$. The uncorrelated scattering (US) property is based on the assumption that the attenuation and phase of a propagation path at delay time τ_i is uncorrelated to the attenuation and phase at delay time τ_k, for $i \neq k$.

In order to apply the concept of the WSSUS channel to real radio channels, the quasi-WSSUS channel (QWSSUS) was introduced by Bello [5]. A QWSSUS channel has the properties of a WSSUS channel within a local area and for a limited bandwidth and time.

Furthermore, it should be noted that for Gaussian processes, the WSS property implies stationarity in the strict sense. If the distribution of the TF $H(f,t)$ is complex Gaussian with zero or nonzero mean, then the amplitude distribution is Rayleigh or Rician, respectively. As this agrees with the channel properties derived in Section 4.2.2, and as Gaussian processes generally simplify any stochastic mathematical analysis, the complex Gaussian case will be assumed.

Figure 4.2 gives an overview of the most commonly used correlation functions and power spectra defining the stochastic properties of the time-variant channel IR and TF. These system functions are found in the center of the figure, surrounded by their second-order moments, which are interrelated by FTs. As mentioned above, our focus lies on the spaced-frequency, spaced-time correlation function depicted just above the center of this figure.

4.3.1.1 Special Cases

Most of the analysis presented in this chapter concentrates on the case of the time-invariant frequency-selective channel. The channel is then described by the TF $H(f)$,

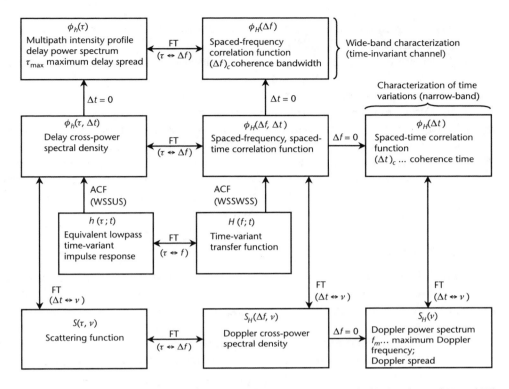

Figure 4.2 Overview of the two time-variant system functions described (the channel IR and TF) and a set of correlation functions (second-order moments) describing their stochastic properties.

which is a WSS complex Gaussian stochastic process in f, according to the above assumptions. The second-order statistical functions characterizing $H(f)$ are the spaced-frequency correlation function $\phi_H(\Delta f) = \phi_H(\Delta f, 0)$ and its FT, the DPS $\phi_h(\tau)$ (see Figure 4.2). A mathematical description of the DPS will be the basis of the so-called FD channel model.

More familiar is the dual approach of modeling the time variability of a narrow-band channel as a WSS complex Gaussian stochastic process $H(t)$. An example for this method is widely known in the literature as Jakes's fading model [1]. Compared to the FD model, the frequency variable is exchanged with the time variable, and the second-order statistics are the spaced-time correlation function $\phi_H(\Delta t) = \phi_H(0, \Delta t)$ and the Doppler power spectrum $S_H(\nu)$ for Doppler frequency ν, which are a Fourier pair as well (see Figure 4.2).

4.3.1.2 Additional Channel Parameters

Figure 4.2 also introduces some additional channel parameters, which are derived from the correlation functions and power spectra.

Coherence time and bandwidth indicate the ranges (in time and frequency) over which the TF $H(f,t)$ shows significant correlation. They are defined as the time or frequency separations Δt and Δf, where the spaced-time or spaced-frequency correlation functions, respectively, drop below 0.9. (Sometimes 0.5 is used for this threshold.)

Related to the power spectra, the maximum delay spread and the Doppler spread are defined, corresponding to the maximum delay time and frequency components in these spectra.

Often, mathematical relations are given in between these parameters, that is, between the coherence-bandwidth and the (reciprocal of the) maximum delay spread or the RDS, and between the coherence time and the (reciprocal of the) Doppler spread. However, these relations lose significance in the Rician case because the dominant component (leading to the nonzero mean of the Gaussian distribution) causes a constant additive term in the channel correlation functions [15]. Therefore, these relationships should be used with care.

4.3.2 Channel Description

The DPS characterizes the frequency selectivity in the FD channel model. In agreement with measurements reported in [13], the shape of the DPS is defined as shown in Figure 4.3. It is specified by four parameters:

ρ^2: The normalized power of the direct ray;
Π [1/s]: The normalized power density of the constant-level part;
τ_1 [s]: The duration of the constant-level part;
γ [1/s]: The decay exponent of the exponentially decaying part.

Mathematically, the DPS can be written as

$$\phi_h(\tau) = \begin{cases} 0 & \tau < 0 \\ \rho^2 \delta(\tau) & \tau = 0 \\ \Pi & 0 < \tau \leq \tau_1 \\ \Pi e^{-\gamma(\tau - \tau_1)} & \tau > \tau_1 \end{cases} \tag{4.15}$$

In many cases, the number of (free) parameters can be further decreased. The exponentially decaying DPS is a good approximation for most practical channels,

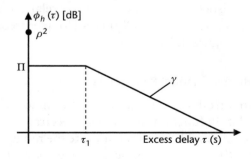

Figure 4.3 Model of the DPS.

which is implemented by letting $\tau_1 = 0$. The existence of an LOS ray at $\tau = 0$ implies that the channel TF is nonzero mean; thus, the fading envelope distribution is Rician. Rayleigh fading channels have $\rho = 0$.

For this analysis, it is appropriate to define $u = \tau_1 \gamma$, which is a single parameter to account for the shape of the DPS. u can take values $u \in [0, \infty]$, where the two extreme cases, $u = 0$ and $u = \infty$, describe an exponentially decaying and a rectangular DPS, respectively. Note that in the latter case (rectangular DPS), the maximum excess delay will be much smaller than for $u = 0$; thus, u can be used to adjust this parameter (see Section 4.3.3).

Relations between the model parameters defined above and the channel parameters are presented in Section 4.3.3.

4.3.3   Relation to (Physical) Channel Parameters

For the application of the FD channel model, it is most important to relate its parameters $\{\rho^2, \Pi, \gamma, \tau_1\}$ to the channel parameters defined in Section 4.2.2: the NRP P_0, the Rician K-factor K, and the RDS τ_{rms}.

The channel parameters derived from the channel model are the local area means, as discussed in Section 4.2. Finite bandwidth realizations or measurements within a local area have "instantaneous" channel parameters $\{\hat{P}_0, \hat{K}, \hat{\tau}_{rms}\}$ spread around those means.

Table 4.1 gives an overview of expressions relating the model parameters $\{\rho^2, \Pi, \gamma, \tau_1\}$ to the channel parameters $\{P_0, K, \tau_{rms}\}$ and vice versa. The derivation of these equations is outlined in Sections 4.3.3.1 through 4.3.3.3. For notational

Table 4.1 Relation Between Model and Channel Parameters (Symbols Are Defined in the Text)

model → channel	
$u = \tau_1 \gamma \in [0, \infty]$	$u = 0$
$P_0 = \rho^2 + \dfrac{\Pi}{\gamma} u_1$	$P_0 = \rho^2 + \dfrac{\Pi}{\gamma}$
$k = \dfrac{\rho^2 \gamma}{\Pi u_1}$	$k = \dfrac{\rho^2 \gamma}{\Pi}$
$\tau_{rms} = \dfrac{1}{\gamma} \sqrt{\dfrac{1}{K+1} \dfrac{u_3}{u_1} - \dfrac{1}{(K+1)^2} \dfrac{u_2^2}{u_1^2}}$	$\tau_{rms} = \dfrac{1}{\gamma} \dfrac{\sqrt{2K+1}}{K+1}$

channel → model	
$u = \tau_1 \gamma$ (must be known)	$u = 0$
$\rho^2 = P_0 + \dfrac{K}{K+1}$	$\rho^2 = P_0 + \dfrac{K}{K+1}$
$\gamma = \dfrac{1}{\tau_{rms}} \sqrt{\dfrac{1}{K+1} \dfrac{u_3}{u_1} - \dfrac{1}{(K+1)^2} \dfrac{u_2^2}{u_1^2}}$	$\gamma = \dfrac{1}{\tau_{rms}} \dfrac{\sqrt{2K+1}}{K+1}$
$\Pi = \dfrac{P_0}{K+1} \cdot \dfrac{\gamma}{u_1}$	$\Pi = \dfrac{P_0}{K+1} \gamma$

convenience we introduce $u_1 = u + 1$, $u_2 = u^2/2 + u = 1$, and $u_3 = u^3/3 + u^2 + 2u + 2$, with $u = \tau_1 \gamma$. An important special case is given by $u = 0$, the exponentially decaying DPS, which is an appropriate description for many practical channels. Table 4.1 also lists the simplified expressions for this case.

4.3.3.1 Derivation of Channel Parameters

From the continuous DPS $\phi_b(\tau)$ defined by (4.15), the analytical expressions given in Table 4.1 can be derived for the expected values of NRP P_0, Rician K-factor K, and RDS τ_{rms}. P_0 relates to the DPS as

$$P_0 = \int_0^\infty \phi_b(\tau)d\tau = \rho^2 + \Pi \cdot \left[\tau_1 + \frac{1}{\gamma}\right] \tag{4.16}$$

The K-factor is used to characterize the amplitude distribution of Rician channels, relating the power of the direct path to the power of the scattered paths.

$$K = \frac{\rho^2}{P_0 - \rho^2} = \frac{\rho^2}{\Pi \cdot (\tau_1 + 1/\gamma)} \tag{4.17}$$

The RDS $_{rms}$ is the single most important parameter characterizing the frequency selectivity. It can be interpreted as the centralized second moment of the normalized DPS

$$\tau_{rms} = \sqrt{\overline{\tau^2} - \left(\overline{\tau}\right)^2} \tag{4.18}$$

where

$$\overline{\tau} = \int_0^\infty \tau \frac{\phi_b(\tau)}{P_0} d\tau = \Pi \cdot \left[\frac{\tau_1^2}{2} + \frac{\tau_1}{\gamma} + \frac{1}{\gamma^2}\right] \tag{4.19}$$

and

$$\overline{\tau^2} = \int_0^\infty \tau^2 \frac{\phi_b(\tau)}{P_0} d\tau = \Pi \cdot \left[\frac{\tau_1^3}{3} + \frac{\tau_1^2}{\gamma} + \frac{2\tau_1}{\gamma^2} + \frac{2}{\gamma^3}\right] \tag{4.20}$$

4.3.3.2 Spaced-Frequency Correlation Function

The spaced-frequency correlation function is used repeatedly throughout this chapter to implement the channel behavior in the mathematical analysis of the radio channel and in the analysis of OFDM system aspects. It is derived from the DPS (4.15) via the FT:

$$\phi_H(\Delta f) = E\{H^*(f)H(f + \Delta f)\} = F\{\phi_b(\tau)\} =$$
$$\rho^2 + \Pi \cdot \tau_1 \mathrm{sinc}(\tau_1 \Delta f)e^{-j\pi\tau_1 \Delta f} + \Pi \cdot \frac{1}{\gamma + j2\pi\Delta f} e^{-j\pi\tau_1 \Delta f} \tag{4.21}$$

For $\tau_1 = 0$, that is, for the special case of an exponentially decaying DPS, the spaced-frequency correlation function can be written as

$$\phi_H(\Delta f) = \frac{P_0}{K+1}\left(K + \frac{1}{1 + j2\pi\Delta f\tau_{rms}K_1}\right) \tag{4.22}$$

where $K_1 = (K+1)/\sqrt{2K+1}$.

4.3.3.3 Maximum Excess Delay

The shape factor u introduces another degree of freedom into the channel model, which allows the variation of the maximum excess delay τ_{max} by a certain factor for a given RDS τ_{rms}. Strictly speaking, the maximum delay spread is infinite due to the exponentially decaying part of the DPS, which never becomes zero. In practice, however, multipath components can be neglected that are attenuated very significantly. We therefore define the maximum excess delay as the delay time, where the exponentially decaying part has decreased by about 43 dB. Such attenuation is reached if the duration of the exponentially decaying part is exactly $\tau_{exp} = 10/\gamma$, leading to the maximum delay spread $\tau_{max} = \tau_1 = \tau_{exp} = \tau_1 + 10/\gamma$. Expressed in terms of channel parameters, this is

$$\tau_{max} = \tau_{rms}(u+10)\frac{u_1(K+1)}{\sqrt{u_1u_3(K+1) - u_2^2}} \tag{4.23}$$

which simplifies for $u = 0$ (i.e., $\tau_1 = 0$) to

$$\tau_{max} = \tau_{rms}10\frac{K+1}{\sqrt{2K+1}} = 10\tau_{rms}K_1 \tag{4.24}$$

It is seen that τ_{max} and τ_{rms} are related by a factor, which is a function of K and u. Figure 4.4 illustrates this factor. According to this definition, τ_{max} is exactly 10 times larger than τ_{rms} at $K = 0$ and $u = 0$. Larger K-factors generally increase this factor; larger parameters u decrease it. For instance, τ_{max} is only about three times τ_{rms} for the rectangular DPS at low K-factors.

As the maximum delay spread τ_{max} defines the maximum "frequency" component of the DPS, it is this parameter that defines the Nyquist frequency when a sampled version of the channel TF is needed in measurements or computer simulations; that is, the sampling interval in the FD must be smaller than $1/(2\tau_{max})$.

4.4 FD Channel Simulation

The discussion of simulation schemes in this section is restricted to the case of static (time-invariant) frequency-selective channels. Such simulations are, for instance, appropriate for the study of OFDM systems, with a system model that reduces the channel, including the inverse FFT (IFFT) at the OFDM transmitter and the FFT at

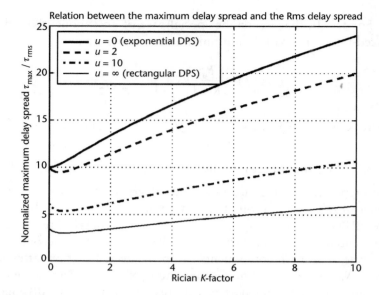

Figure 4.4 Factor between the maximum delay spread τ_{max} and the RDS τ_{rms}, as a function of the Rician factor K and with the shape factor u as a parameter.

the receiver to a set of parallel Gaussian (sub-) channels (see Chapter 5, [17]). These subchannels have complex attenuation factors given by the channel's TF $H(f)$ at the frequency instants of the OFDM SCs $f = nF$, where F [Hz] is the sampling interval in the FD, and $n = \{0, 1, 2, ..., N - 1\}$. The simulation scheme presented in this section directly generates $H(f)$ for well-defined channel parameters.

In some cases, for instance, when evaluating channel-estimation schemes, the time variability of the TF is also of great importance. The extension of the static simulation scheme to a time-variant one is discussed.

4.4.1 Model Description

The simulation system for time-invariant channels is shown in Figure 4.5. Real-valued white (or wideband) Gaussian random processes $W(f)$ in the FD are generated by a noise source. The appropriate spaced-frequency correlation is obtained by FD filtering of $W(f)$ with a (lowpass) filter $g(f)$. The output of this filter is the real-valued, colored noise process $_rH'(f) = W(f) * g(f)$, where * denotes convolution. The inverse FT of $_rH'(f)$ (in delay-TD representation) is complex valued and hermitian, that is, symmetric with respect to the $\tau = 0$ axis. It is not causal, in contrast to the IR of a real channel. The required causality in the TD is obtained by applying the Hilbert transform (HT) to $_rH'(f)$ and adding the result $_iH'(f)$ as $H'(f) = _rH'(f) - j_iH'(f)$. Doing this, the negative part of the IR is canceled.

The amplitude of the TF $|H'(f)|$ is Rayleigh distributed since $H'(f)$ is a complex Gaussian noise process. A Rician fading channel may be simulated by adding a complex constant $\rho \cdot e^{j\theta_\rho}$ to $H'(f)$, representing the LOS path at $\tau = 0$.

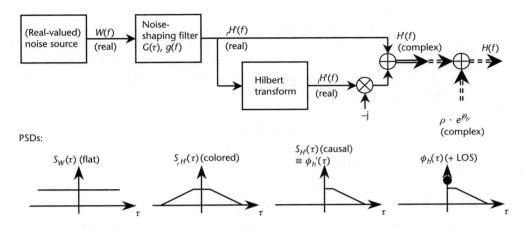

Figure 4.5 FD simulation of the frequency-selective radio channel.

4.4.2 Implementation of the Simulation Scheme

To obtain a computer simulation program producing TFs with the desired DPS, two elements of the above simulation scheme must be appropriately designed; the noise-shaping filter $g(f)$ and the variance of the noise source σ_W^2. The simulator produces a sampled version of the TF, $H(nF)$, where F [Hz] is the sampling interval in the FD and $n = \{0, 1, 2, ..., N - 1\}$. F must be selected according to the sampling theorem, that is, $F < 1/(2\tau_{max})$.

The power spectral density (PSD) of the output of the simulation scheme (which is in τ domain) has to match the continuous DPS defined by (4.15). This is achieved by designing the filter $g(nF)$ to have a TF $G(\tau)$ proportional to the DPS (for $\tau > 0$, that is, skipping the LOS component). Any classic filter design method can be used in this process [11]. By definition we let $|G(\tau)| = 1$ during the constant-level part (or at $\tau = 0^+$ if there is no constant-level part), which leads to the variance $\sigma_W^2 = \Pi/(4F)$, as derived next.

4.4.2.1 Derivation of the Variance of the Noise Source

The noise source produces independent, real-valued noise samples with variance σ_W^2. The sequence $W(nF)$ thus has a (periodic) spectrum with constant PSD

$$S_W(\tau) = \sigma_W^2 F \tag{4.25}$$

Applying these samples to the noise-shaping filter with amplitude TF

$$|G(\tau)| = \begin{cases} 1 & |\tau| \le \tau_1 \\ e^{-\gamma(|\tau|-\tau_1)} & |\tau| > \tau_1 \end{cases} \tag{4.26}$$

leads to the PSD of $\mathrm{Re}\{H'(nF)\}$ written as

$$S_{,H'}(\tau) = \sigma_W^2 F |G(\tau)| \tag{4.27}$$

The next step in the simulation scheme is the addition of the Hilbert-transformed sequence, which increases the PSD for $\tau > 0$ by a factor of four. (The HT cancels the negative τ part of the Fourier spectrum, while doubling the positive τ part, resulting in fourfold power for $\tau > 0$). This yields the PSD to be compared with the model (the DPS) as

$$S_{,H'}(\tau) = 4\sigma_W^2 \, F|G(\tau)| \equiv \phi_h(\tau) \quad \text{for } \tau > 0 \tag{4.28}$$

which yields $\sigma_W^2 = \Pi/(4F)$.

4.4.2.2   Extension to a Time-Variant Channel Simulator

In order to extend this static simulation scheme to a time-variant one, the TF $H(f_1,t)$ must have the required Doppler spectrum when the time variations are investigated at any given frequency $f = f_1$. This may be achieved by generating a number of independent TFs $H(f,t = k \cdot T_s)$, $k = \{1, 2, 3, \ldots\}$ and filtering them in time direction at each frequency sample, according to a specific Doppler spectrum. [Separability of the joint time-frequency correlation function $\phi_H(\Delta f, \Delta t)$ is thereby assumed.[4]] A set of N filters is required for applying time variability to the TFs in this way. It should be noted that this simulation scheme gets rather complex. It might thus be preferable to use a conventional fading simulator, one that generates a (time-variant) IR, and transform the IR to the FD, if required. Usually, the IR is defined by much fewer than N coefficients; therefore, the complexity is reduced.

4.4.3   FD Simulation Results

Figure 4.6(a) shows a simulated TF, which is compared to a measured one in Figure 4.6(b). The two channels' IRs are given in Figure 4.7(a, b), both derived from the respective TFs using the IDFT without windowing. The measurement was performed with a network analyzer, observing a bandwidth of 1 GHz around a center frequency of 11.5 GHz.[5] The channel parameters $P_0 = 62.1$ dB, $K = 1.9$ dB, and $\tau_{rms} = 9.0$ ns were extracted from the measured TF [Figure 4.6(b)] and (with $\tau_1 = 0$) used to generate the simulated TF [Figure 4.6(a)]. Both TFs have a length of 801 samples. A 15-tap FIR filter was used for the noise-shaping filter $g(nF)$ in the simulation scheme.

The TF is obtained from a stochastic simulation model. Therefore, we do not expect it to be identical to the measured TF. However, it is clearly seen that the characteristic of the fading is well reproduced. Originally, a linearly increasing phase

---

4.   Separability of the two-dimensional spaced-frequency, spaced-time correlation function $\phi_H(f,t)$ means that it can be written as a product $\phi_H(\Delta f, \Delta t) = \phi_H(\Delta f) \cdot \phi_H(\Delta t)$. This assumption is valid if $\tau_{max} f_m \ll 1$ [18], which is given for practical propagation channels (τ_{max} denotes the maximum excess delay; f_m is the Doppler spread).

5.   The author would like to thank Dr. G. J. M. Janssen for providing measurement results for the validation of the proposed methods [14]. The measurements were conducted at the TNO Physics and Electronics Laboratory in The Hague, the Netherlands, between August and December 1991.

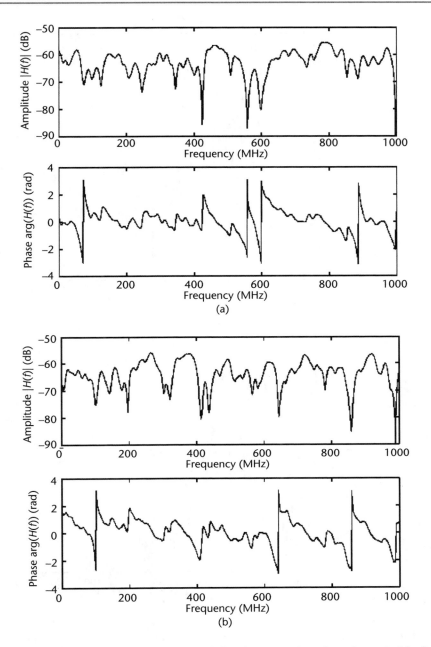

Figure 4.6   (a) Simulated TF, and (b) TF measured with a network analyzer (corrected for linear phase shift).

shift was evident in the measured TF corresponding to the propagation delay of the shortest path. In the illustration this was compensated for to have the first component arrive at (excess) delay $\tau = 0$ in agreement with the simulation model.

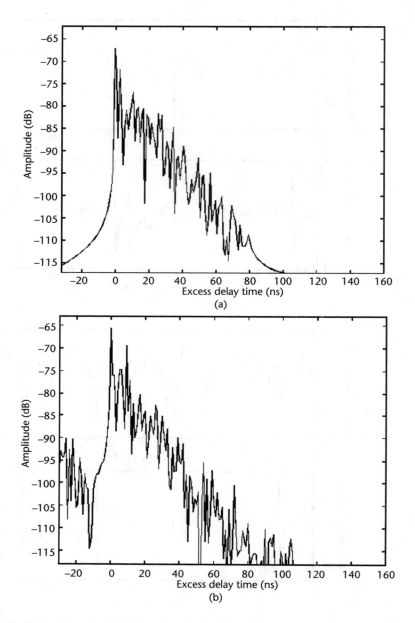

Figure 4.7 (a) IR derived by IDFT from the simulated TF shown in Figure 4.6(a), and (b) IRs derived from the measured TF [see Figure 4.6(b)].

The PDF and the cumulative distribution function (CDF) of the simulated amplitude TF are shown in Figure 4.8 and compared to the Rayleigh distribution. Because of the low K-factor ($K = -1.9$ dB), good agreement is evident.

Figure 4.9 shows second-order statistical properties estimated from simulated TFs. The power spectrum obtained by averaging periodograms of 100 simulated

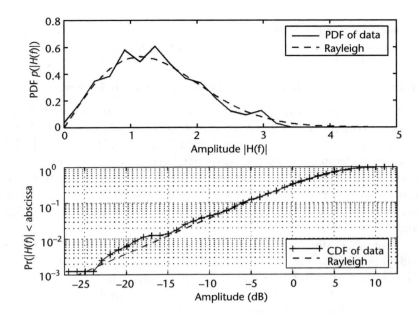

Figure 4.8   PDF and CDF of the amplitude of the simulated TF.

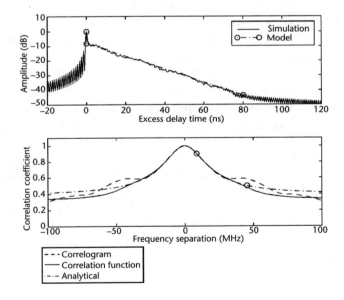

Figure 4.9   DPS and spaced-frequency correlation function for the FD simulation model. Upper plot: DPS and estimated power spectrum. Lower plot: Correlogram; estimated and analytical correlation functions; markers "O"; coherence bandwidths 0.5 and 0.9.

TFs agrees well with the used DPS model (upper plot). The lower plot shows (spaced-frequency) correlation properties and compares them to the theoretical function given by (4.21). Coherence bandwidths are determined by solving numerically for the frequency-separations, where the correlation function's magnitude drops to 0.9 (or 0.5, according to the definition).

4.4.4 Differences from Time-Domain Simulation Schemes

The equally spaced tap gains of the sampled (and band-limited) IR are not uncorrelated, according to the analysis shown in Section 4.2.2 [see (4.12)]. That is, there exists a certain autocorrelation between different delay bins of the IR. This correlation results from the band limitation needed for the time quantization, which implies a convolution of the discrete, nonsampled IR with a sinc function. Another effect of this convolution is visible in spectral components at negative delay values due to leakage effects. The FD simulation scheme shows these properties [see Figure 4.7(a) and Figure 4.9].

Many TD simulators, however, implement the channel IR by simply generating independent, complex-valued path gains at the (sample-spaced) delay bins [18, 19]. Leakage effects, that is, components at negative delays, are not considered either (see, e.g., [6], Figure 1.12; [19]). (One sampled simulation model, which does consider those effects, is described in [18].)

In particular, these simplifications are used, when the channel models are applied for the design of digital radio interfaces. Normally, the resulting differences are negligible, but there are cases where the impact gets important. An example is DFT-based channel-estimation schemes for OFDM. Such channel estimators determine first a coarse estimate of the channel TF, for instance from a training sequence. In order to reduce the mean-square-error (MSE) of the estimate, the next step is a transform of the TF to the delay TD, yielding a noisy channel IR. In this form, likely noise components, at negative or very large delay values, can be identified and set to zero, followed by a back transformation to the FD. The result is an estimate of the channel TF with enhanced SNR and hopefully reduced MSE. The simulation of such a scheme suggests excellent performance, if a so-called sample-spaced channel simulator is used because then the channel IR is indeed zero at the sample bins set to zero. On a real channel, however, important information is lost, as channel taps are set to zero, which correspond to leakage components. This leads to irreducible error floors in terms of bit error rate (BER) and MSE [20–22]. In this respect, the proposed FD simulation model has an inherent advantage over conventional, sampled TD models because the correlation among channel taps and leakage effects is considered in closer agreement to reality.

4.5 Application to Millimeter-Wave Radio Channels

This section has two main purposes. First, the suitability of the proposed FD channel model and simulation scheme is verified; second, parameters are found for the model. These goals are approached through a discussion of measurement campaigns reported in the literature.

In particular, our focus lies on mm-wave radio channels. Within the millimeter-wave frequencies, the 60-GHz band has received most attention in the literature, mainly for the following reasons: Large amounts of bandwidth are unallocated in this band, bandwidths that are required for communications systems at the intended data rates of 100 Mbps and above. Another advantage of the 60-GHz band is due to a physical property of the propagation channel at this specific frequency. Oxygen absorption leads to attenuation above 11 dB/km between 57 and 63 GHz. This attenuation (in addition to the path loss) is believed to enable shorter reuse distances in cellular systems because it counteracts cochannel interference. Over short distances, the additional attenuation can be neglected.

It is a general property of millimeter-wave propagation that the behavior of propagation rays is well characterized by geometric optics. That is, waves do not penetrate through walls or other obstacles, and wave reflection is the main mechanism leading to multipath. Scattering, diffraction, and wave guiding are considered far less important [23].

This section starts with a discussion of measurement results (Section 4.5.1). In Section 4.5.2, typical channel parameter values are given. The influence of features of the propagation environment on those parameters is discussed. Channel models suggested in the literature are treated separately in Section 4.5.3. Section 4.5.4 deals with the applicability of the newly proposed FD channel model to (millimeter-wave) radio channels—the validation of the FD model.

4.5.1   Discussion of Measurement Results

A major activity in the field of millimeter-wave propagation has been conducted in the framework of the Research into Advanced Communications systems in Europe project 2067, Mobile Broadband Systems (RACE-MBS) [23, 24]. The measurement campaigns described include material characterization and indoor and outdoor propagation studies. Ray-tracing models have been developed for predicting propagation parameters and for investigating the impact of environment features, antenna characteristics, and the like. An extensive list of literature on millimeter-wave propagation is found in the "Final Report on Propagation Aspects" of the RACE-MBS project [23].

Partly related is the activity carried out within the European Cooperation in the Field of Scientific and Technical Research (COST 231) program [8]. This study also covers indoor and outdoor channels. A major contribution to indoor propagation originates from the research of P. F. M. Smulders, conducted at Eindhoven University of Technology [13].

Other work on indoor channels is found in [12, 14, 15, 25–35]; outdoor studies are presented in [36–38]. Note that most of the work has been done on indoor channels and their modeling, probably because of the range limitation of millimeter-wave propagation.

The main parameters of interest for applying the FD model to millimeter-wave channels are the NRP P_0, the Rician K-factor K, and the RDS τ_{rms}. For the air interface design, the latter two parameters, $\{K, \tau_{rms}\}$, are generally sufficient. The NRP is required for link budget considerations.

While most studies present results of τ_{rms} and NRP, the K-factor is unfortunately commonly not investigated.

Generally, it is difficult to compare measurements conducted by various research groups because of the following:

- Differences in the measurement equipment and method used;
- Different antenna characteristics and configurations;
- Different parameters measured and presented;
- Different environments investigated.

We try to organize this comparison and overview by discussing the parameters of interest and elaborating on the impact of some of the above listed factors. Only wideband measurements are considered because of the importance of characterizing the time-dispersive and frequency-selective nature of the channel. The modeling of these channel properties is essential for the air interface design, which is the intended application of the channel model under development.

4.5.1.1   Measurement Setups and Techniques

Most indoor-measurement campaigns use vector network analyzers to scan the channel TF (phase and magnitude) versus frequency (see, e.g., [8, 12–15, 31, 33–35]). The conditions to use such equipment are short distances because a phase reference must be provided between the transmitting and the receiving sides and a (quasi-) static channel due to the time it takes to acquire the frequency TF. These conditions are feasible in indoor scenarios. The main advantages of this approach are high time resolution achieved by scanning over a large bandwidth and good SNR because a narrowband (continuous wave) signal is transmitted in which the whole transmit power is concentrated. The delay-time resolution investigated is normally around 1 ns, corresponding to a scanning bandwidth of 1 to 2 GHz. In [38], a network analyzer was used for outdoor measurements.

Correlation-type channel sounders were developed for the extensive measurement campaigns performed in the RACE-MBS project [23, 24]. For outdoor channels, a wideband test signal (chirp) was generated by rapidly sweeping a carrier over a bandwidth of up to 200 MHz [36, 39]. A separate indoor channel sounder is based on the transmission of a pseudorandom binary sequence and a sliding correlator on the receiver's side (see [40]). Similar equipment was employed in [32] for indoor measurements and in [37] for outdoor measurements.

Within this Ph.D. research, a novel, noncoherent channel measurement technique was developed that can estimate the NRP, K-factor, and τ_{rms} from swept-frequency power measurements. No phase measurement is required, which simplifies the equipment needed. Measurement campaigns conducted with this method at DUT are described in [25–28]. Indoor and outdoor channels were studied at 17 and 60 GHz.

4.5.2   Discussion of Channel Parameters

The RDS τ_{rms} and the Rician K-factor are the two most important parameters for specifying the channel's frequency-selective nature in the context of air interface design. It will be seen from the study of the FD level crossing rate that the RDS determines the number of fades per bandwidth, while the K-factor specifies the depth of the fades. The NRP just determines the average SNR. Since the SNR is usually considered a variable in any kind of system studies, absolute values of NRP are not of major importance for the air interface design.

4.5.2.1   RDS

The following main features of the propagation environment influence the RDS. Note that the mentioned properties are applicable for indoor channels only. Similar features, however, will also have an impact on the RDS in outdoor scenarios.

Room Size
Generally, the RDS increases with the room size. Such behavior was, for instance, reported in the work of Smulders [13], who measured typical values of RDS between 15 and 45 ns in small rooms with dimensions $\leq 24 \times 11 \times 4.5$ m^3 and values between 30 and 70 ns in larger rooms.

These values are rather large, compared to the results from many other indoor-measurement campaigns found in the literature. The main reason for the large values is, next to the large rooms investigated, the antenna design implemented. The biconical horn antennas, having an omnidirectional radiation pattern in the azimuth plane, a 3-dB beam width of 9° in the elevation plane, and a directivity of 9 dBi, were designed such that the NRP hardly depends on the antenna's position within a room. Therefore, they radiate quite a large fraction of the transmitted power toward the walls, leading to strong first reflections and long delay spreads. Although the delay spreads are quite large, this design might be of advantage because self-shadowing effects become less harmful. That is, signal loss due to the obstruction of the LOS path by the user (see [41]) is assumed to be less significant for such antenna setups.

Antenna Directivity
Directive antennas attenuate parts of the impinging reflected waves. Therefore, the RDS usually decreases, as more directive antennas (in the azimuth plane) are employed.

Such behavior is clearly seen from measurements and ray-tracing simulations performed by T. Manabe et al. [31]. In a room with dimensions of $13.5 \times 8 \times 2.6$ m^3, they measured typical RDS values of 18, 14, 5, and 1 ns for, respectively, an omnidirectional antenna ($\lambda/2$-dipol) and antennas with 3-dB beam widths of ~60°, ~10°, and ~5°.

A similar study based on a ray-tracing tool is presented in the final report of the RACE-MBS project [23]. In the investigated room of approximate dimensions $11 \times 7 \times 3$ m^3, different antenna configurations were evaluated, leading to RDS values of 20 to 25 ns for the less directional antennas and values (significantly) lower than 5 ns for the most directional ones.

In order to investigate this anticipated dependency between the RDS and the antenna characteristics, Smulders has conducted some additional measurements using a 15-dBi circular-horn antenna (in stead of the 9-dBi omnidirectional biconical horn antenna) on one side of the measured link [13]. His results confirm the expected behavior if median values of RDS are considered (RDS decreases from ~40 to ~25 ns). However, the maximum RDS values observed were even larger than for the standard antenna configuration (increase from ~48 to ~60 ns). Bultitude et al. [33], who performed a measurement campaign at 40 GHz in a large open office environment, reported a similar behavior. It is a possible explanation that the more directive antenna, which also has higher gain, may emphasize some reflected paths with a rather large delay time. Such paths contribute strongly to the RDS.

Building Material

The reflectivity of building material is expected to be another important factor influencing the RDS. This behavior was reported, for example, by Smulders [13], who measured higher RDS values in a small room with metal walls (room dimensions ~10 × 9 × 3 m^3; $\tau_{rms} \equiv 45$ ns) than in a much larger auditorium room with walls covered by wood and acoustically soft material (room dimensions ~30 × 21 × 6 m^3; $\tau_{rms} \equiv 35$ ns). In a small room (~13 × 9 × 4 m^3) with wood-covered walls, RDS values of ~20 ns were measured.

Outdoor Measurements

Measurements in seven different streets in downtown Oslo were reported in [23, 24, 36] (MBS-RACE project). The RDS is typically lower than 20 ns, except for one measurement where a major reflection source (tourist bus) was located on the street. In the latter case the RDS was less than 50 ns. The maximum delay spread (the sliding delay window, or the shortest period of the IR containing 90% of the received energy) is less than 45 ns for 90% of the measurement points. However, maximum values up to 270 ns were observed. Results from measurements at city squares show higher values of RDS and maximum delay spread.

Outdoor measurement results reported in [25, 26] (for 17 GHz), [38] (for 60 GHz), and [37] (for 40 and 60 GHz) also show RDS values starting below 20 ns and occasionally reaching about 100 ns and above [38]. Little work has been done on outdoor propagation in millimeter-wave bands.

4.5.2.2   Rician K-Factor

Parameter pairs of RDS, τ_{rms}, and the Rician K-factor are required for modeling multipath radio channels using the FD channel model introduced in Section 4.3. While statistics of the RDS are found in most propagation studies, the Rician K-factor is often not (explicitly) investigated. Many studies assume Rayleigh fading amplitude distributions, that is, K-factors of zero. In situations where the LOS between transmitter and receiver is obstructed, this assumption may be reasonable. However, as an LOS path is often required for reliable transmission at millimeter-wave frequencies [41], the K-factor becomes an important channel parameter. Typical values of K are given below. The influence of an LOS path and the influence of the antenna characteristics are investigated.

Influence of an LOS Path

Janssen [14, 15] and Bohdanowicz [25, 26] conducted two measurement campaigns that consider the K-factor. Although those measurements were performed at lower frequency bands (at 2.4, 4.75, and 11 GHz, and at 17 GHz, respectively), the results are interesting for modeling the 60-GHz channel. One important reason is that most of the measurement situations were similar to the expected scenarios for 60-GHz systems, where both the transmitter and the receiver are typically located within the same room. Moreover, results for all of these frequency bands are quite similar, suggesting that a shift to the 60-GHz band would not have a large impact either. The comparative study of a 1.7-GHz and a 60-GHz channel presented in [32] confirms the latter.

Characteristic channel parameters reported by Janssen ([14, 15]) are $\tau_{rms} \equiv 10$ ns, $K \equiv 2.5$ dB in LOS situations, and τ_{rms} 15 ns, $K \equiv 3$ dB without LOS. All measurements were performed in relatively small rooms.

The 17-GHz channel study by Bohdanowicz ([25, 26]) gives typical K-factors between 0.3 and 2.5 dB for LOS indoor scenarios ($\tau_{rms} \cong 5$ to 17 ns) and values around 1 dB for indoor non-LOS situations ($\tau_{rms} \cong 9$ ns). Larger K-factors of 3 to 5 dB were determined from outdoor LOS measurements, where $\tau_{rms} \cong 20 \ldots 30$ ns.

Clearly, increased K-factors are observed in the presence of a (dominant) LOS path, corresponding to amplitude distributions with shallower fades. While K-factors below 3 dB can be well represented by the Rayleigh distribution, higher values should be modeled by the Rician distribution. In particular, if $K \geq 0$ dB, that is, if the dominant path carries greater or equal power than all of the reflected paths, the Rician model must be used.

Antenna Directivity

It is expected that more directive antennas yield higher K-factors because if the antennas are pointed toward one another, the dominant path is amplified while the reflected ones are attenuated. Inspection of the IRs shown by Manabe et al. in [31] confirm such behavior. Unfortunately, no values of K-factors are given there.

The channel model parameters given by Smulders [13] and by Kunisch et al. [12] can be used to estimate the Rician K-factor and investigate the impact of antennas' directivity. Those model parameters were obtained from 60-GHz channel measurements.

Smulders's model parameters [13] imply that even for the 15-dBi directive antenna and in the presence of an LOS path, the K-factor would be less than or equal to 6 dB and therefore well described by the Rayleigh model. Note that in this study the directive (receive) antenna was not pointed toward the transmitter. This may be a partial explanation for this unexpected result.

Kunisch's model parameters [12] correspond to Rician K-factors between 7.3 dB and 25 dB (and τ_{rms} between 5.7 and 1 ns, respectively). Kunisch's measurement setup used an 8-dBi antenna at the transmitter and two receiver antennas pointed toward the transmitter, with respective gains of 20 and 22 dBi. It appears that such an antenna configuration can effectively reduce the multipath fading. Adaptive antennas (beam forming) can avoid the need to point the antenna manually.

Larger K-factors reduce τ_{rms} when the decay exponents of the average PDP remain constant. This is also seen from the equations given in Table 4.1, where γ should be considered constant. The model parameters given by Kunisch [12] confirm that such a dependency may exist, at least within one room.

4.5.3 Overview of Channel Models

Most of the (stochastic) channel models proposed in the literature for millimeter-wave channels are based on the indoor propagation model presented by Saleh and Valenzuela [16]. First, this section reviews their model. Second, a number of modifications are discussed for its application to millimeter-wave channels. The suitability of the FD channel model for this frequency band is studied in Section 4.5.4.

4.5.3.1   Review of the Saleh and Valenzuela Model

The Saleh and Valenzuela model is a method to generate time-discrete channel IRs as defined by (4.1). Stochastic processes are specified to model the ray arrival times $\{\tau_i\}$, the ray amplitudes $\{\beta_i\}$, and the ray phases $\{\theta_i\}$.

The ray phases are considered to be independent RVs that are uniformly distributed over $[0, 2\pi)$ because the phases vary over that range when the path lengths change by just one wavelength.

Two Poisson processes implement the ray-arrival process. Reflections are assumed to arrive in clusters, where the first Poisson process models the arrival times of the clusters with some fixed rate Λ [1/s].[6] Subsequent ray arrivals within the clusters are realized by the second Poisson process with rate $\lambda_\tau \gg \Lambda$. Per definition, the first ray and the first cluster arrive at $\tau = 0$. A Poisson process of (ray) arrivals implies exponentially distributed interarrival times, written as

$$p(\Delta\tau) = \lambda_\tau \exp\left[-\lambda_\tau(\Delta\tau)\right] \tag{4.29}$$

where $\Delta\tau$ is the delay-time difference between consecutive paths of the same cluster.

The probability distribution of the path gains $\{\beta_i\}$ is a Rayleigh distribution. (Therefore, the path gains including the uniformly distributed path phases $\left\{\beta_i e^{j\theta_i}\right\}$ follow a complex Gaussian distribution.) Introducing the variables l and k for indexing the cluster and ray-within-cluster, respectively, the mean square values of the magnitudes $\left\{\beta_{kl}^2\right\}$ are written

---

6. According to [16], the formation of clusters is related to the building superstructure; that is, clusters of rays typically originate from (steel-reinforced) exterior or interior walls or large metal doors or objects. The rays within the clusters are due to reflections in the vicinity of the transmitter or receiver. Clustering of rays is therefore a property of indoor channels at longer ranges and at lower carrier frequencies, where propagation through walls is possible. Clustering of rays also occurs in outdoor channels [42–44].

$$\overline{\beta_{kl}^2} = \overline{\beta_{00}^2} e^{-T_l/\Gamma} e^{-\tau_{kl}/\gamma_{sv}} \tag{4.30}$$

where $\{T_l\}$ and $\{\tau_{kl}\}$ are the cluster and ray-within-cluster arrival times, respectively, and Γ and γ_{sv} are the corresponding power decay time constants. This function is called the average PDP because it characterizes the average ray power of the IR as a function of the excess delay time. It is composed of a set of exponentially decaying parts, one for each cluster of rays.

For more details on the Saleh and Valenzuela model, the reader is referred to [16].

4.5.3.2   Modifications to the Model

Several authors have applied a number of modifications to the above-described model in order to match it to millimeter-wave channels.

Most of the implementations found in the literature reduce the number of clusters to one (see, e.g., [12–15]). This simplification is made because in a typical indoor millimeter-wave channel, the reflections originate all from within one room, leading to a single, dense cluster of ray arrivals. Remember that millimeter-wave frequencies hardly penetrate through building material. An exception is the work of Park [35], who gives a set of parameters for the original multicluster version of Saleh and Valenzuela's model. Park investigated indoor channels at 60 GHz.

In several cases, the model has been augmented by a separately specified path at $\tau_0 = 0$ ([12–15]) in order to extend the model to Rician channels by introducing a (dominant) LOS path.

Smulders [13] proposes a composite average PDP, where the exponentially decaying part of the single cluster is preceded by a constant-level part. This part is introduced to better describe first-order reflections arriving at similar strength due to the antenna design chosen. Such a constant-level part is also implemented in the FD channel model proposed in Section 4.3.2.

Janssen shows in [15] how to adjust the generated discrete-time IRs in order to realize exactly a given set of channel parameters $\{P_0, K, \tau_{rms}\}$. (Note that those are local area mean parameters; see Section 4.2.2.) He also suggests a method to incorporate small-scale fading effects resulting from movements within a local area. That mechanism is based on ray-arrival directions relative to the assumed transceiver movement.

4.5.4   Applicability of the FD Model

The proposed FD channel model characterizes the mobile radio channel by its DPS, the FT of the spaced-frequency correlation function. This model agrees well with the modified (single-cluster) versions of the Saleh and Valenzuela model introduced above, because the DPS of the FD model is described in an almost equivalent way to the average PDP of the (single-cluster) Saleh and Valenzuela model. In the FD model, Rician channels may be implemented using the discrete, direct path at $\tau_0 = 0$. Moreover, a constant-level part is incorporated as in [13], which allows for a better match to certain channel IRs and also enables varying the maximum excess delay in some range, as investigated in Section 4.3.3.

According to the overview of channel models presented above, a single cluster of rays is an appropriate description of millimeter-wave indoor channels, where the transmitter and receiver are typically located within the same room. The suitability of the proposed channel model for such scenarios was also confirmed by the comparison of measurement results to computer simulations (see Section 4.4.3).

Clustering of rays can be implemented in the proposed model by modifying the DPS accordingly, that is, by defining a DPS consisting of multiple exponentially decaying parts. Similarly, arbitrary outdoor channels could be realized. However, using the model for the design of OFDM systems, I am confident that the simple model introduced in Section 4.3.2 is applicable to a much wider range of actual environments. For such systems, the most important channel properties are the correlation among (adjacent) SCs and the distribution of their amplitudes (or powers). These properties are well preserved by the model as long as the channel parameters are matched to the environment under investigation. And these parameters can be freely (and easily) chosen in the proposed model. It can be even proven that for Rayleigh fading channels and for small frequency separations, the correlation coefficient in FD is independent of the channel model. (The correlation coefficient is the normalized autocorrelation function of the squared magnitudes of the TF.) This finding strongly supports the claim that such a simple stochastic model is sufficient for many air interface design problems.

4.6   Conclusions

The main novelties discussed in this chapter concern the so-called FD channel model and its implementation on a computer simulation scheme. The FD model is the FD dual of Jakes's Doppler-spectrum model [1, 3, 4]. Just as in Jakes's model, the (narrowband) channel's time variability is described by the spaced-time correlation function and by the Doppler spectrum; in the FD model the (time-invariant) channel's frequency selectivity is described by the spaced-frequency correlation function and by the delay power spectrum. (The power spectra and correlation functions are interrelated by FTs.) The simulation scheme introduced directly generates realizations of channel TFs with well-defined channel parameters. Note that a frequency-selective channel is equivalent to a time-dispersive (multipath) channel. The major advantages of the proposed models are as follows:

- Good agreement with physical propagation channels, in particular in millimeter-wave frequency bands and in indoor environments (see Section 4.5);
- Availability of analytical expressions relating model parameters to physical channel parameters and vice versa, allowing straightforward matching of the model to any given environment (see Section 4.3.3);
- Suitability for OFDM system design, which is the goal of this book (see Chapter 7);

- Simplicity of the model, which allows for the mathematical analysis of many aspects of transmission schemes, like the performance evaluation and optimization of BERs, synchronization, and channel-estimation schemes (see Chapter 7);
- Availability of an efficient simulation model (see Section 4.4).

However, the extension of the simulation model from the static version presented to a time-variant version is rather complex. This may be a disadvantage of the FD model.

The (physical) channel parameters specifying the FD model are elaborately discussed. The channel at a local area of dimensions of a few wavelengths (approximately 5 to 40 λ) is defined by a set of fixed parameters: the normalized (or average) received power P_0, the Rician K-factor K, and the RDS τ_{rms}. At a limited observation bandwidth, however, these parameters appear to be time (or location) variant within a local area because individual propagation paths are not resolvable and multipath interference between them leads to rapid (small-scale) variations of the resulting channel IR. Reduced-bandwidth simulations performed with the FD model also show a variability of these instantaneous parameters among realizations.

It was suggested that the RDS τ_{rms} and the Rician K-factor are equally important for the characterization of frequency-selective multipath radio channels. In the next chapter, it will be shown that τ_{rms} effectively specifies the number of fades per bandwidth and their average bandwidth, while the K-factor describes the depth of fades. Most experimental studies, however, investigate τ_{rms} only. The K-factor is analyzed in rather few cases, although LOS conditions and directional antennas are commonly considered, two factors that are anticipated to increase the K-factor. Rayleigh fading channels have a K-factor of zero.

Channel parameters depend on a number of features of the propagation environment and the antenna setup. Larger rooms and more reflective building materials generally increase the RDS τ_{rms}. Higher antenna directivity decreases τ_{rms} and increases the K-factor. The presence of an LOS between the transmit and receive antennas leads to larger K and sometimes to lower τ_{rms}. Within the same environments and with similar antenna setups, the frequency band has surprisingly little influence on those parameters. Table 4.2 lists typical parameter values with short descriptions of the main features of their corresponding environments. Most investigations of these channels conclude that an LOS between the transmitter and the receiver is required for reliable communications. However, the results from [13], where a special antenna design was used, suggest that the reflections can be sufficient as well.

Since the channel parameters are influenced by many factors and in ways that are hard to predict, a method is desirable to measure them cheaply and simply. The next chapter presents a method that can be used to estimate these parameters $\{P_0, K, \tau_{rms}\}$ accurately from scans of the channel's power response versus frequency. Standard laboratory equipment can be used to apply that scheme.

Table 4.2 Typical Channel Parameters of Frequency-Selective Millimeter-Wave Radio Channels

Title	Comments and Reference	Antenna Configuration	RDS (ns)	Rician K-factor (dB)
Small/medium room, LOS	Measurement at 2.4, 4.75, and 11 GHz [14, 15]	~2.5-dBi biconical antennas (~100° beam width)	10	2.5
Small/medium rooms, non-LOS			15	3 (Rayleigh)
Outdoor	17 GHz [25, 26]		30	3
Medium room, directional antenna	High-gain antenna pointed at BS; 60 GHz [12]	BS: 8 dBi PS: ~20 dBi	5	10
Computer room (~10 × 9 × 3 m³)	60 GHz [13]	9-dBi biconical antennas (~9° beamwidth)	45	0 (Rayleigh)
Large hall (~43 × 41 × 7 m³)			60	0 (Rayleigh)
Corridor (~45 × 2.5 × 3 m³)			75	0 (Rayleigh)
Lecture room (~13 × 9 × 4 m³)			20	0 (Rayleigh)

BS: base station; PS: portable station.

References

[1]   Jakes, W. C., Jr., *Microwave Mobile Communications*, New York: Wiley-Interscience, 1974.

[2]   Proakis, J. G., *Digital Communications*, 3rd ed., New York: McGraw-Hill, 1995.

[3]   Gans, M. J., "A Power-Spectral Theory of Propagation in the Mobile-Radio Environment," *IEEE Trans. Veh. Technol.*, Vol. VT-21, No. 1, February 1972, pp. 27–38.

[4]   Clarke, R. H., "A Statistical Theory of Mobile-Radio Reception," *Bell Syst. Tech. J.*, Vol. 47, July–August 1968, pp. 957–1000.

[5]   Bello, P. A., "Characterization of Randomly Time-Variant Linear Channels," *IEEE Trans. on Commun. Systems*, Vol. CS-11, December 1963, pp. 360–393.

[6]   Steele, R., *Mobile Radio Communications*, New York: John Wiley & Sons, 1992.

[7]   Rappaport, T. S., *Wireless Communications: Principles and Practice*, Upper Saddle River, NJ: Prentice Hall, 1996.

[8]   European Commission, Cost Action 231, *Digital Mobile Radio towards Future Generation Systems*, Final Report, EUR 18957, Luxembourg, 1999.

[9]   Rice, S. O., "Mathematical Analysis of Random Noise," *Bell Syst. Tech. J.*, Vol. 23, July 1944, pp. 282–332, and Vol. 24, January 1945, pp. 46–156.

[10] Rice, S. O., "Statistical Properties of a Sine Wave Plus Random Noise," *Bell Syst. Tech. J.*, Vol. 27, 1948, pp. 109–157.

[11] Oppenheim, A. V., and R. W. Schafer, *Discrete-Time Signal Processing*, 2nd ed., Upper Saddle River, NJ: Prentice Hall, 1999.

[12] Kunisch, J., et al., "MEDIAN 60 GHz Wideband Indoor Radio Channel Measurements and Model," *Proc. IEEE Vehic. Techn. Conf. (VTC'99–Fall)*, Amsterdam, the Netherlands, September 1999, pp. 2393–2397.

[13] Smulders, P. F. M., "Broadband Wireless LANs: A Feasibility Study," Ph.D. thesis, Eindhoven University of Technology, Eindhoven, the Netherlands, 1995.

[14] Janssen, G. J. M., P. A. Stigter, and R. Prasad, "Wideband Indoor Channel Measurements and BER Analysis of Frequency Selective Multipath Channels at 2.4, 4.75 and 11.5 GHz," *IEEE Trans. on Communications,* Vol. 44, No. 10, October 1996, pp. 1272–1288.

[15] Janssen, G. J. M., "Robust Receiver Techniques for Interference-Limited Radio Channels," Ph.D. thesis, Delft University of Technology, Delft, the Netherlands, June 1998.

[16] Saleh, A. A. M., and R. A. Valenzuela, "A Statistical Model for Indoor Multipath Propagation," *IEEE J. Select. Areas Commun.* Vol. 5, No. 2, February 1987, pp. 128–137.

[17] Edfors, O., et al., *An Introduction to Orthogonal Frequency-Division Multiplexing, Division of Signal Processing,* Luleå University of Technology, Research Report TULEA 1996:16, at www.sm.luth.se/csee/sp/publications.html.

[18] Zhang, W., "Simulation and Modelling of Multipath Mobile Channels," *Proc. VTC'94 (IEEE Vehicular Technology Conference),* Stockholm, Sweden, 1994, pp. 160–164.

[19] Pahlavan, K., and A. H. Levesque, *Wireless Information Networks,* New York: John Wiley & Sons, 1995.

[20] Chini, A., M. S. Tanany, and S. A. Mahmoud, "Transmission of High Rate ATM Packets over Indoor Radio Channels," *IEEE J. Select. Areas in Communications,* Vol. 14, No. 3, April 1996, pp. 469–476.

[21] Edfors, O., "Low-Complexity Algorithms in Digital Receivers," Ph.D. thesis, Luleå University of Technology, Luleå, Sweden, September 1996.

[22] van de J. J., et al., "On Channel Estimation in OFDM Systems," *Proc. IEEE Vehic. Technol. Conf.,* Chicago, IL, July 1995, pp. 815–819.

[23] Correia, L. M., et al., *Final Report on Propagation Aspects,* RACE 2067, Deliverable R2067/IST/2.2.5/DS/P/070.b1, RACE Central Office, European Commission, Brussels, Belgium, December 1995.

[24] Mohamed, S. A., et al., *Report on Propagation Measurements,* RACE 2067, Deliverable R2067/BTL/2.2.2/DS/P/035.b1, RACE Central Office, European Commission, Brussels, Belgium, December 1994.

[25] Bohdanowicz, A., G. J. M. Janssen, and S. Pietrzyk, "Wideband Indoor and Outdoor Multipath Channel Measurements at 17 GHz," *Proc. VTC'99–Fall (IEEE Vehicular Technology Conference),* Amsterdam, the Netherlands, September 1999, pp. 1998–2003.

[26] Bohdanowicz, A., *Wideband Indoor and Outdoor Radio Channel Measurements at 17 GHz,* Delft University of Technology, UbiCom-Technical Report/2000/2, January 2000, at http://www.ubicom.tudelft.nl/docs.

[27] El Hattachi, R., et al., "Characterization and Simulation of the 18 GHz Radio Channel," *Proc. IEEE Benelux 6th Symposium on Vehicular Technology and Communications,* Brussels, Belgium, October 1998.

[28] Purwaha, J., et al., "Wideband Channel Measurements at 60 GHz in Indoor Environments," *Proc. IEEE Benelux 6th Symposium on Vehicular Technology and Communications,* Brussels, Belgium, October 1998.

[29] Fernandes, J. J. G., J. C. Neves, and P. F. M. Smulders, "MM-Wave Indoor Radio Channel Modelling vs. Measurements," *Wireless Personal Communications,* Vol. 1, No. 3, 1995, pp. 211–219.

[30] Kato, A., et al., "Measurements of Millimeter Wave Indoor Propagation and High-Speed Digital Transmission Characteristics at 60 GHz," *Proc. PIMRC'97 (IEEE 8th Intern. Symp. on Personal Indoor Mobile Radio Commun.),* Helsinki, Finland, September 1997, 149–154.

[31] Manabe, T., Y. Miura, and T. Ihara, "Effects of Antenna Directivity on Indoor Multipath Propagation Characteristics at 60 GHz," *Proc. PIMRC'95 (IEEE 6th Intern. Symp. on Personal Indoor Mobile Radio Commun.),* September 1995, pp. 1035–1039.

[32] Davies, R., et al., "Wireless Propagation Measurements in Indoor Multipath Environments at 1.7 GHz and 60 GHz for Small Cell Systems," *Proc. 41st IEEE Veh. Techn. Conf.*, St. Louis, MS, May 1991, pp. 589–593.

[33] Bultitude, R. J. C., R. F. Hahn, and R. J. Davies, "Propagation Considerations for the Design of an Indoor Broad-Band Communications System at EHF," *IEEE Trans. Veh. Technology*, Vol. 47, No. 1, February 1998, pp. 235–245.

[34] Hübner, J., et al., "Simple Channel Model for 60 GHz Indoor Wireless LAN Design Based on Complex Wideband Measurements," *Proc. IEEE Vehic. Techn. Conf. (VTC'97)*, 1997, pp. 1004–1008.

[35] Park, J. H., et al., "Analysis of 60 GHz Band Indoor Wireless Channels with Channel Configurations," *Proc. PIMRC'98 (IEEE 9th Intern. Symp. on Personal Indoor Mobile Radio Commun.)*, Taiwan, September 1998, pp. 617–620.

[36] Løvnes, G., J. J. Reis, and R. H. Rækken, "Channel Sounding Measurements at 59 GHz in City Streets," *Proc. PIMRC'94 (IEEE 5th International Symposium on Personal Indoor Mobile Radio Communications)*, The Hague, the Netherlands, September 1994, pp. 496–500.

[37] Wales, S. W., and D. C. Rickard, "Wideband Propagation Measurements of Short Range Millimetric Radio Channels," *Electronics and Commun. Eng. Journal*, August 1993, pp. 249–254.

[38] Daniele, N., D. Chagnot, and C. Fort, "Outdoor Millimetre-Wave Propagation Measurements with Line of Sight Obstructed by Natural Elements," *IEE Electronics Letters*, Vol. 30, No. 18, September 1994, pp. 1533–1534.

[39] Løvnes, G., S. E. Paulsen, and R. H. Rækken, "A Versatile Channel Sounder for Millimetre Wave Measurements," *Proc. PIMRC'93 (IEEE 4th International Symposium on Personal Indoor Mobile Radio Communications)*, Yokohama, Japan, September 1993.

[40] Cox, D. C., "Delay Doppler Characteristics of Multipath Propagation at 910 MHz in a Suburban Mobile Radio Environment," *IEEE Trans. Antennas and Propagation*, Vol. 20, No. 5, September 1972, pp. 625–635.

[41] Flament, M., "On 60 GHz Wireless Communication Systems," Ph.D. thesis, Chalmers University of Technology, Göteborg, Sweden, 2000.

[42] Turin, G. L., et al., "A Statistical Model of Urban Multipath Propagation," *IEEE Trans. Veh. Technology*, Vol. VT-21, February 1972, pp. 1–9.

[43] Suzuki, H., "A Statistical Model for Urban Radio Propagation," *IEEE Trans. Communications*, Vol. COM-25, July 1977, pp. 673–680.

[44] Hashemi, H., "Simulation of the Urban Radio Propagation Channel," *IEEE Trans. Veh. Technology*, Vol. VT-28, August 1979, pp. 213–224.

CHAPTER 5

Basics of OFDM and Synchronization

5.1 Introduction

The aim of this chapter is to provide some theoretical background on the OFDM transmission technique. A brief introduction to OFDM is given in Section 5.2. We review the block diagram of a "classic" OFDM system, which employs a GI to mitigate the impairments of the multipath radio channel. We also discuss several design considerations related to hardware properties and derive the mathematical model for an idealized system, leading to the conclusion that data symbols can be transmitted independently of each other [i.e., without ISI and intercarrier interference (ICI)]. Moreover, the effects of synchronization imperfections are analyzed, like carrier-frequency and phase offsets and timing errors.

Section 5.3 introduces a method of calculating uncoded BERs for this idealized OFDM system model. This method is largely based on work presented in [1]. Differential and coherent detection schemes can be evaluated for Rayleigh and Rician fading channels. We also show that, for the system proposal under investigation, differential detection in time direction is much preferable to differential detection in frequency direction. Imperfect synchronization and channel estimation may be assessed by extending the system model used and by incorporating the SNR degradations due to ICI and ISI. Basic aspects are discussed in this chapter. Issues for further refinement of the methods are addressed.

Section 5.4 presents conclusions and recommendations.

5.2 OFDM Introduction and System Model

OFDM is a parallel transmission scheme, where a high-rate serial data stream is split up into a set of low-rate substreams, each of which is modulated on a separate SC (FDM). Thereby, the bandwidth of the SCs becomes small compared with the coherence bandwidth of the channel; that is, the individual SCs experience flat fading, which allows for simple equalization. This implies that the symbol period of the substreams is made long compared to the delay spread of the time-dispersive radio channel.

By selecting a special set of (orthogonal) carrier frequencies, high spectral efficiency is obtained because the spectra of the SCs overlap, while mutual influence among the SCs can be avoided. The derivation of the system model shows that by introducing a cyclic prefix (the GI), the orthogonality can be maintained over a dispersive channel (see Section 5.2.3).

This section starts with a brief introduction to the OFDM transmission technique, based on the description of the system's block diagram. We then discuss some hardware-related design considerations (Section 5.2.2) that become relevant if an OFDM system is implemented in hardware. For instance the dc SC and the SCs near the Nyquist frequency must be avoided. Next, we derive the system model for a perfectly synchronized system (Section 5.2.3), and we investigate the impact of the most relevant synchronization errors (Section 5.2.4).

For a more elaborate introduction to OFDM, the reader may refer to the relevant chapters of [2, 3] and to [4–6]. An excellent overview of the effects of many nonideal transmission conditions is given in [7], wherein numerous further references are found.

5.2.1 OFDM Introduction and Block Diagram

Figure 5.1 shows the block diagram of a simplex point-to-point transmission system using OFDM and FEC coding. The three main principles incorporated are as follows:

1. The IDFT and the DFT are used for, respectively, modulating and demodulating the data constellations on the orthogonal SCs [8]. These signal-processing algorithms replace the banks of I/Q-modulators and demodulators that would otherwise be required. An analysis of Section 5.2.3 will show this equivalence.

    Note that at the input of the IDFT, N data constellation points $\{x_{i,k}\}$ are present, where N is the number of DFT points. (i is an index on the SC; k is an index on the OFDM symbol). These constellations can be taken according to any phase shift keying (PSK) or QAM signaling set (symbol mapping). The N output samples of the IDFT, being in TD, form the baseband signal carrying the data symbols on a set of N orthogonal SCs. In a real system, however, not all of these N possible SCs can be used for data, as Section 5.2.2.3 discusses.

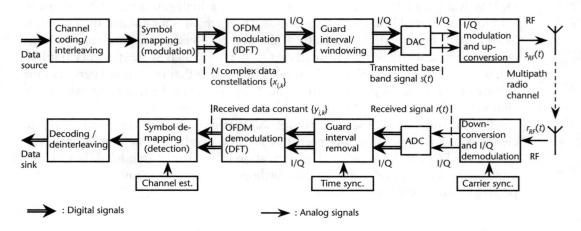

Figure 5.1   Simplex point-to-point transmission using OFDM.

Usually, N is taken as an integer to the power of two, enabling the application of the highly efficient (inverse) FFT algorithms for modulation and demodulation.

2. The second key principle is the introduction of a cyclic prefix as a GI, whose length should exceed the maximum excess delay of the multipath propagation channel [9]. Due to the cyclic prefix, the transmitted signal becomes periodic, and the effect of the time-dispersive multipath channel becomes equivalent to a cyclic convolution, discarding the GI at the receiver. Due to the properties of the cyclic convolution, the effect of the multipath channel is limited to a pointwise multiplication of the transmitted data constellations by the channel TF, or the FT of the channel IR; that is, the SCs remain orthogonal (see [4–7]). This conclusion will also follow from the derivation of the system model in Section 5.2.3. The only drawback of this principle is a slight loss of effective transmit power, as the redundant GI must be transmitted. Usually, the GI is selected to have a length of one tenth to a quarter of the symbol period, leading to an SNR loss of 0.5 to 1 dB (see also Figure 5.2).

The equalization (symbol demapping) required for detecting the data constellations is an elementwise multiplication of the DFT output by the inverse of the estimated channel TF (channel estimation). For phase modulation schemes, multiplication by the complex conjugate of the channel estimate can do the equalization. Differential detection can be applied as

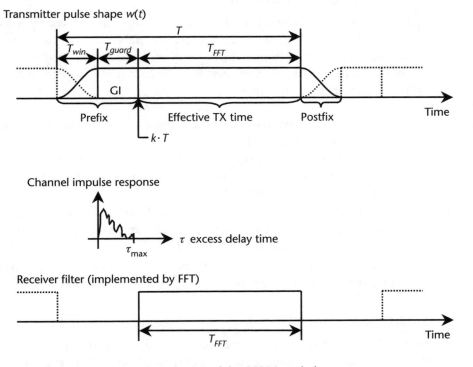

Figure 5.2   Cyclic extension and windowing of the OFDM symbol.

well, where the symbol constellations of adjacent SCs or subsequent OFDM symbols are compared to recover the data.

3. FEC coding and (FD) interleaving are the third crucial idea applied. The frequency-selective radio channel may severely attenuate the data symbols transmitted on one or several SCs, leading to bit errors. Spreading the coded bits over the bandwidth of the transmitted system, an efficient coding scheme can correct for the erroneous bits and thereby exploit the wideband channel's frequency diversity. OFDM systems utilizing error-correction coding are often referred as coded OFDM (COFDM) systems. The BER of the uncoded system is analyzed in Section 5.3.

The complex equivalent baseband signals generated by digital signal processing are in-phase/quadrature (I/Q)–modulated and up-converted to be transmitted via an RF carrier. The reverse steps are performed by the receiver.

Synchronization is a key issue in the design of a robust OFDM receiver. Time and frequency synchronization are paramount, respectively, to identify the start of the OFDM symbol and to align the modulators' and the demodulators' local oscillator frequencies. If any of these synchronization tasks is not performed with sufficient accuracy, then the orthogonality of the SCs is (partly) lost. That is, ISI and ICI are introduced. The effect of small synchronization errors is analyzed in Section 5.2.4.

5.2.2 Design of the OFDM Signal

The proposal of a realistic OFDM-based communications system was one of the goals of this research project. Therefore, we elaborate here on some hardware-related design considerations, which are often neglected in theoretical studies. Elements of the transmission chain that have impact on the design of the transmitted OFDM signal include the following:

- The time-dispersive nature of the mobile channel. The transmission scheme must be able to cope with this.
- The bandwidth limitation of the channel. The signal should occupy as little bandwidth as possible and introduce a minimum amount of interference to systems on adjacent channels.
- The TF of the transmitter/receiver hardware. This TF reduces the useable bandwidth compared to the theoretical one given by the sampling theorem. That is, some oversampling is required.
- Phase jitter and frequency offsets of the up- and down-converters, and Doppler spreading of the channel.

5.2.2.1 GI

As mentioned earlier, a GI is introduced to preserve the orthogonality of the SCs and the independence of subsequent OFDM symbols, when the OFDM signal is transmitted over a multipath radio channel. The GI, a cyclic prefix, is a copy of the last part of the OFDM symbol, which is transmitted before the so-called effective part of the symbol (see Figure 5.2). Its duration T_{guard} is simply selected to be larger than the maximum excess delay of the (worst-case) radio channel. Therefore, the effective part of the received signal can be seen as the cyclic convolution of the transmitted OFDM symbol by the channel IR.

5.2.2.2   Windowing

A rectangular pulse has a very large bandwidth due to the sidelobes of its FT being a sinc function. Windowing is a well-known technique to reduce the level of these sidelobes and thereby reduce the signal power transmitted out of band. In an OFDM system, the applied window must not influence the signal during its effective period. Therefore, cyclically extended parts of the symbol are pulse shaped as depicted in Figure 5.2 [3].

Note that this additional cyclic prefix extends the GI to some extent; that is, the delay-spread robustness is slightly enhanced. On the other hand, the efficiency is further reduced, as the window part is also discarded by the receiver. The orthogonality of the SCs of the OFDM signal is restored by the rectangular receiver filter implemented by the DFT (Figure 5.2), requiring the correct estimation of the DFT start time kT, where T is the OFDM symbol period.

The symbol periods in Figure 5.2 are given as times. Since the implementation is usually done on digital hardware, those periods are also often defined in terms of samples. N, N_{guard}, and N_{win} then define the number of samples in the effective part, guard, and windowing interval, respectively. The effective part is also referred to as the FFT part because this part of the OFDM symbol is applied to the FFT to recover the data at the receiver.

Spectrum of the Transmitter Pulse Shape
Windowing of the transmitter pulse using a raised-cosine function can be seen as a convolution of the extended rectangular pulse of duration T with a sine half-wave, as shown in Figure 5.3. In the FD, this convolution means a multiplication of the sinc spectrum of the rectangular pulse with the spectrum of the sine half-wave. It is seen that this multiplication reduces the sidelobes of the transmitter pulse shape.

In Figure 5.3(a), the zeros of the spectrum occur at positions $i \cdot F = i/T_{FFT}$, $i = \{\pm 1, \pm 2, \ldots\}$, that is, at those positions where the adjacent SCs are located. The extension of the rectangular pulse to length $T = T_{FFT} + T_{guard} + T_{win}$ reduces the distance between zeros to $1/T$ [Figure 5.3(b)]. The windowing function [Figure 5.3(c)] has zeros at positions $\pm 1/T_{win}\{3/2, 5/2, 7/2, \ldots\}$.

5.2.2.3   System TF (ADCs, DACs, IF-Filters, RF Front Ends)

Because of the lowpass filters required for the analog-to-digital (ADC) and digital-to-analog (DAC) conversion of the transmitted and received (baseband) signals, not all N SCs can be used, if an N-point IFFT is applied for modulation. The SCs close to the Nyquist frequency $f_s/2$ will be attenuated by these filters and, thus, cannot be

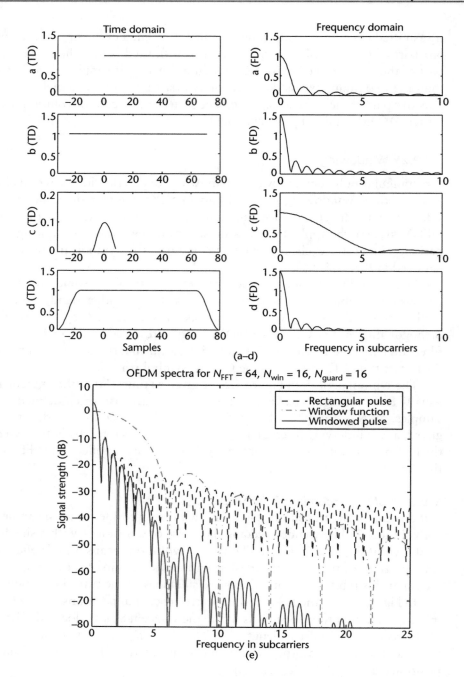

Figure 5.3 (a) Shape and spectrum of the OFDM receive filter (realized by FFT); (b) rectangular pulse of duration T and its spectrum; (c) sine half-wave used for pulse shaping and its spectrum; (d) transmitter pulse prototype $w(t)$ and its spectrum; (e) spectra of (b)–(d) in logarithmic scale.

used for data transmission (see Figure 5.4). ($f_s=1/T_s$ is the sampling frequency.) Also the dc SC might be heavily distorted by dc offsets of the ADCs and DACs, by carrier feedthrough, and so forth, and should thus be avoided for data.

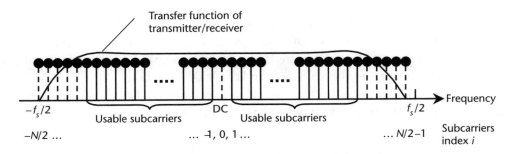

Figure 5.4   TF of the transmitter/receiver hardware and its impact on the design of an OFDM system.

5.2.3   OFDM System Model

The above-introduced features of the OFDM signal are defined mathematically in this section. This will lead to the conclusion that, using the OFDM principle, data symbols can be transmitted over multipath radio channels without influencing each other.

5.2.3.1   Signal Model and Definitions

Mathematically, the OFDM signal is expressed as a sum of the prototype pulses shifted in the time and frequency directions and multiplied by the data symbols. In continuous-time notation, the kth OFDM symbol is written

$$
s_{RF,k}(t-kT) = \begin{cases} \mathrm{Re}\left\{ w(t-kT) \displaystyle\sum_{i=-N/2}^{N/2-1} x_{i,k} e^{\,j2\pi\left(f_c+\frac{i}{T_{FFT}}\right)(t-kT)} \right\} \\ kT - T_{win} - T_{guard} \le t \le kT + T_{FFT} + T_{win}\ \textit{otherwise} \\ 0 \end{cases} \tag{5.1}
$$

Most of the mathematical symbols have been defined in the previous figures already. A complete list of symbols is given here:

T: Symbol length; time between two consecutive OFDM symbols;

T_{FFT}:   FFT time; effective part of the OFDM symbol;

T_{guard}:   GI; duration of the cyclic prefix;

T_{win}:   Window interval; duration of windowed prefix/postfix for spectral shaping;

f_c: Center frequency of the occupied frequency spectrum;

$F = 1/T_{FFT}$:   Frequency spacing between adjacent SCs;

N: FFT length; number of FFT points;

k: Index on transmitted symbol;

i: Index on SC; $i \in \{-N/2, -N/2+1, \ldots -1, 0, 1, \ldots N/2-1\}$;

$x_{i,k}$:   Signal constellation point; complex {data, pilot, null} symbol modulated on the ith SC of the kth OFDM symbol;

$w(t)$: The transmitter pulse shape defined as

$$w(t) = \begin{cases} \frac{1}{2}\left[1 - \cos \pi\left(t + T_{win} + T_{guard}\right)/T_{win}\right] & -T_{win} - T_{guard} \leq t < -T_{guard} \\ 1 & -T_{guard} \leq t \leq T_{FFT} \\ \frac{1}{2}\left[1 - \cos \pi\left(t - T_{FFT}\right)/T_{win}\right] & T_{FFT} < t \leq T_{FFT} + T_{win} \end{cases} \quad (5.2)$$

Finally, a continuous sequence of transmitted OFDM symbols is expressed as

$$s_{RF}(t) = \sum_{k=-\infty}^{\infty} S_{RF,k}(t - kT) \quad (5.3)$$

The simulated spectrum of such an OFDM signal is depicted in Figure 5.5 for different window lengths.

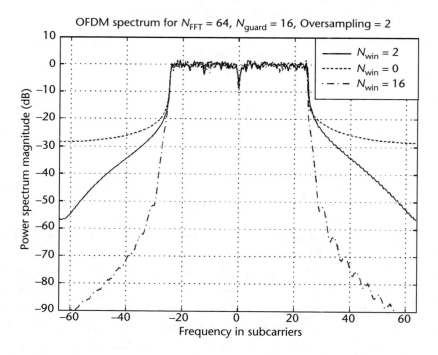

Figure 5.5 Spectrum of an OFDM signal with 64 SCs and different window lengths. Twofold oversampling has been applied in the TD; 48 SCs are used for data.

5.2.3.2   Lowpass Equivalent Transmitted Signal

From (5.1) to (5.3), the complex equivalent lowpass signal transmitted can be directly given. The complex envelope of the OFDM signal is written

$$s(t) = \sum_{k=-\infty}^{\infty} S_k(t - kT) \tag{5.4}$$

with

$$s_k(t - kT) = \begin{cases} w(t - kT) \sum_{i=-N/2}^{N/2-1} x_{i,k} e^{j2\pi\left(\frac{i}{T_{FFT}}\right)(t - kT)} \\ kT - T_{win} - T_{guard} \leq t \leq kT + T_{FFT} + T_{win} \ otherwise \\ 0 \end{cases} \tag{5.5}$$

Note the similarities of this expression to the equation of a Fourier series

$$v(t) = \sum_{n=-\infty}^{\infty} c(nf_0) e^{j2\pi nf_0 t} \tag{5.6}$$

where the complex-valued Fourier coefficients $c(nf_0)$ represent the complex-valued signal constellation points $x_{i,k}$, and the frequencies nf_0 correspond to the SC frequencies i/T_{FFT}.

In a digital system, this modulated waveform can be generated by an IDFT or by its computationally efficient implementation, the IFFT. The data constellations $x_{i,k}$ are the input to this IFFT; the TD OFDM symbol is its output.

5.2.3.3   Time-Dispersive Channel

The influence of the time-variant, multipath fading radio channel is expressed by its (lowpass equivalent) IR $h(\tau,t)$, plus additive white Gaussian noise (AWGN) $n(t)$:

$$r(t) = h(\tau,t) * s(t) + n(t) = \int_0^{\tau_{max}} h(\tau,t)s(t - \tau)d\tau + n(t) \tag{5.7}$$

The range of integration in this convolutional integral (* denotes convolution) has been limited to $[0, \tau_{max}]$ because the channel IR is zero elsewhere. Excess delay $\tau = 0$ of the channel is defined as the delay time at which the first wave arrives at the receiver. Thus, transmit and receive time instants are mathematically defined as equal (Figure 5.2). τ_{max} is the maximum excess delay of the channel.

Two assumptions are made to simplify the derivation of the received signal. The channel is considered quasistatic during the transmission of the kth OFDM symbol; thus, $h(\tau,t)$ simplifies to $h_k(\tau)$. Furthermore, we define the maximum excess delay

$\tau_{max} < T_{guard}$. Therefore, there is no interference of one OFDM symbol on the effective period of the consecutive one (see Figure 5.2); that is, ISI is suppressed in case of sufficiently accurate time synchronization.

5.2.3.4   OFDM Demodulation

The demodulation of the OFDM signal should be performed by a bank of filters, which are "matched" to the effective part $[kT, kT+T_{FFT}]$ of the OFDM symbol (see Figure 5.2). The reverse operation to (5.6), that is, the extraction of the Fourier coefficients $c(nf_0)(=x_{i,k})$ from the TD signal $v(t)(=r(t))$, exactly formulates such a bank of matched filters. It is written

$$c(nf_0) = \frac{1}{T_0} \int_{T_0} v(t)e^{-j2\pi nf_0 t}\, dt \tag{5.8}$$

where T_0 is the integration period equivalent to T_{FFT}. In a digital implementation, a DFT or (preferably) an FFT is used to realize these filters.

Assuming knowledge of the exact time instants kT at which the OFDM symbols start, we try to extract the transmitted signal constellations $x_{i,k}$ from the received signal $r(t)$. The received signal constellations are denoted $y_{i,k}$.

$$\begin{aligned}
y_{i,k} &= \frac{1}{T_{FFT}} \int_{t=kT}^{kT+T_{FFT}} r(t)e^{-j2\pi i(t-kT)/T_{FFT}}\, dt = \\
&\quad \frac{1}{T_{FFT}} \int_{t=kT}^{kT+T_{FFT}} \left[\int_{\tau=0}^{\tau_{max}} h_k(\tau)s(t-\tau)d\tau + n(t) \right] e^{-j2\pi i(t-kT)/T_{FFT}}\, dt
\end{aligned} \tag{5.9}$$

Because of the integration ranges in (5.9) and $\tau_{max} < T_{guard}$, there is no influence on the adjacent OFDM symbols transmitted, and $s(t)$ can be replaced by $s_k(t)$ [see (5.5)].

$$\begin{aligned}
y_{i,k} &= \frac{1}{T_{FFT}} \int_{t=kT}^{kT+T_{FFT}} \left[\int_{\tau=0}^{\tau_{max}} h_k(\tau) \sum_{i'=-N/2}^{N/2-1} x_{i',k} e^{j2\pi\left(\frac{i'}{T_{FFT}}\right)(t-kT-\tau)} d\tau \right] e^{-j2\pi i(t-kT)/T_{FFT}}\, dt + \\
&\quad \frac{1}{T_{FFT}} \int_{t=kT}^{kT+T_{FFT}} n(t)e^{-j2\pi i(t-kT)/T_{FFT}}\, dt
\end{aligned} \tag{5.10}$$

Note that $w(t-kT) = 1$ in the range of integration. The window is thus omitted in this equation. The second integral in (5.10) leads to independent additive noise samples $n_{i,k}$ since the complex exponential terms represent orthogonal functions. Substituting $u = t - kT$ for the ease of notation and changing the order of integration and summation yields

$$\begin{aligned}
y_{i,k} &= \sum_{i'=-N/2}^{N/2-1} x_{i',k} \frac{1}{T_{FFT}} \int_{u=0}^{T_{FFT}} \left[\int_{\tau=0}^{\tau_{max}} h_k(\tau)e^{-j2\pi i'(u-\tau)/T_{FFT}} d\tau \right] e^{-j2\pi iu/T_{FFT}}\, du + n_{i,k} = \\
&\quad \sum_{i'=-N/2}^{N/2-1} x_{i',k} \frac{1}{T_{FFT}} \int_{u=0}^{T_{FFT}} \left[\int_{\tau=0}^{\tau_{max}} h_k(\tau)e^{-j2\pi i'\tau/T_{FFT}} d\tau \right] e^{-j2\pi(i-i')u/T_{FFT}}\, du + n_{i,k}
\end{aligned} \tag{5.11}$$

The inner integral of the second expression represents the FT of $h_k(\tau)$ at the frequency instants $i'/T_{FFT} = i'F$, which is the sampled channel TF at time kT. It is expressed by the channel coefficients

$$h_{i',k} = FT\{h_k(\tau)\} = \int_{\tau=0}^{\tau_{max}} h_k(\tau)e^{-j2\pi i'\tau/T_{FFT}}\, d\tau = H(i'F, kT) \tag{5.12}$$

Using this notation, the output of the receiver filter bank simplifies to

$$y_{i,k} = \sum_{i'=-N/2}^{N/2-1} x_{i',k} h_{i',k} \frac{1}{T_{FFT}} \int_{u=0}^{T_{FFT}} e^{-j2\pi(i-i')u/T_{FFT}}\, du + n_{i,k} \tag{5.13}$$

The integral in this equation has the value 1, only if $i = i'$. For $i \neq i'$, with i and i' being integer values, the integral is zero. Thus, we finally obtain

$$y_{i,k} = x_{i,k} h_{i,k} + n_{i,k} \tag{5.14}$$

From this form it is seen that a perfectly synchronized OFDM system can be viewed as a set of parallel Gaussian channels as depicted in Figure 5.6 [4–6]. The multipath channel introduces an attenuation/amplification and phase rotation according to the (complex-valued) channel coefficients $\{h_{i,k}\}$.

Channel estimation is required in order to retrieve the data contained in these signal constellations because the receiver must have a phase (and amplitude) reference to detect the transmitted symbol correctly. Differential detection can be used alternatively, in which case the decision is made by comparing the phases (and amplitudes) of symbols transmitted over adjacent SCs or subsequent OFDM symbols.

Due to the attenuation/amplification, each SC typically has an individual SNR. The SNR per SC (after the DFT) is defined as

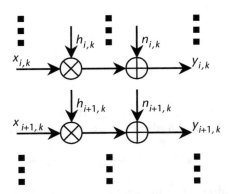

Figure 5.6 Idealized OFDM system model. The SCs of the OFDM system can be considered as parallel Gaussian channels under the assumptions of perfect time and carrier synchronization and perfect suppression of multipath by the GI.

$$(E_c/N_0)_{i,k} = E\left\{|x_{i,k}|^2\right\}|h_{i,k}|^2 / \sigma_N^2 \tag{5.15}$$

where $\sigma_N^2 = E\left\{|n_{i,k}|^2\right\}$ is the noise variance. With the NRP being written $P_0 = E\left\{|h_{i,k}|^2\right\}$, the average SNR becomes $\overline{E_c/N_0} = E\left\{|x_{i,k}|^2\right\}P_0 / \sigma_N^2$. Usually, the signal energy is normalized to unity; that is, $E\left\{|x_{i,k}|^2\right\} = 1$.

5.2.4 Synchronization Errors

As an introduction to the work on synchronization algorithms, this section reviews the relevant effects of synchronization errors. Original work on this topic is found in numerous publications (see, e.g., [10, 11]). A comprehensive overview is given in [7].

5.2.4.1 FFT Time Synchronization Error

The impact of an FFT timing offset at the receiver can be analyzed mathematically by shifting the integration interval of the matched filter bank [see (5.9)]. For a timing error of δt, the ideal interval $t \in [kT, kT+T_{FFT}]$ becomes $t \in [kT+\delta t, kT+T_{FFT}+\delta t]$ and (5.9) is written

$$y_{i,k} = \frac{1}{T_{FFT}} \int_{t=kT=\delta t}^{kT+T_{FFT}+\delta t} r(t)e^{-j2\pi i(t-kT-\delta t)/T_{FFT}} \, dt \tag{5.16}$$

δt is assumed to be sufficiently small (typically $\delta t < T_{guard}$) that no ISI arises due to the timing error. In other words, the error is small enough for the channel IR to remain within the GI. Therefore, the receiver window still does not overlap with the preceding or consecutive OFDM symbol; that is, no energy is collected from these adjacent OFDM symbols, and the demodulated signal can be expressed from the transmitted symbol $s_k(t)$ again [compare (5.10)]. Following the same steps as in Section 5.2.3 [see (5.9)–(5.14)], we obtain for the second part of (5.11) (with $u = t-kT-\delta t$),

$$y_{i,k} = \sum_{i'=-N/2}^{N/2-1} x_{i',k} \frac{1}{T_{FFT}} \int_{u=0}^{T_{FFT}} \left[\int_{\tau=0}^{\tau_{max}} h(\tau)e^{-j2\pi i'\tau/T_{FFT}} \, d\tau\right] e^{-j2\pi[(i-i')u+i'\delta t]/T_{FFT}} \, du + n_{i,k} \tag{5.17}$$

Moving the term $e^{-j2\pi i'\delta t/T_{FFT}}$ out of the integral yields the expression for the demodulated signal constellations in case of a timing error,

$$y_{i,k} = x_{i,k}h_{i,k}e^{-j2\pi i\delta t/T_{FFT}} + n_{i,k} = x_{i,k}h_{i,k}e^{-j2\pi i\delta t/N} + n_{i,k} \tag{5.18}$$

where $\delta t'$ is the timing offset in samples. It is evident that a timing offset gives rise to a progressive phase rotation of the signal constellations. The phase rotation is zero at the center frequency, and it linearly increases toward the edges of the frequency band. It is easily verified from (5.18) that a timing offset in one sample introduces a phase shift of $\pm\pi$ to the outermost SCs (having $i \cong \pm N/2$), regardless of the FFT length. In Figure 5.7, this effect is visualized for a 64-carrier OFDM system with zero carriers at f_c and at the edges of the frequency band.

If coherent detection is utilized, the induced progressive phase rotation is detected implicitly by the channel-estimation algorithm. The subsequent equalization (SC-wise multiplication of the received symbols by the inverse of the estimated channel coefficients) will thus automatically correct for small timing offsets. No performance degradation is thereby caused. However, if the timing offset is too large, ISI and ICI are introduced because energy is also collected from one of the adjacent OFDM symbols, leading to a partial loss of orthogonality [7].

Differential detection is also robust to small timing-offsets. If the differential detection is applied in the frequency direction, the progressive phase rotation may, however, reduce the distance between the compared constellation points, which can lead to performance degradation. Such performance results are given in Section 5.3.3.

A (small) sampling frequency offset leads to a (slowly) increasing timing offset and, therefore, to a progressive phase rotation at an increasing slope. Larger errors yield ICI because the SC-spacing at the receiver can no longer be assumed to equal the SC spacing at the transmitter. (The SC spacing is defined as $F = 1/(NT_s)$, where T_s is the sampling period.)

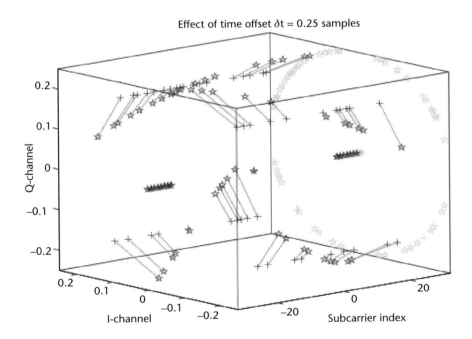

Figure 5.7   Visualization of the influence of an FFT timing offset on the demodulated signal constellations. A linearly increasing phase rotation is observed with increased frequency distance to the center frequency. A "+" indicates a QPSK constellation without the influence of a timing offset; a "☆" depicts a rotated data symbol.

5.2.4.2 Carrier Synchronization Error

Frequency offsets are typically introduced by a (small) frequency mismatch in the local oscillators of the transmitter and the receiver. Doppler shifts can be neglected in indoor environments.

The impact of a frequency error can be seen as an error in the frequency instants, where the received signal is sampled during demodulation by the FFT. Figure 5.8 depicts this twofold effect. The amplitude of the desired SC is reduced ("+"), and ICI arises from the adjacent SCs ("O").

Mathematically, a carrier offset can be accounted for by a frequency shift δf and a phase offset θ in the lowpass equivalent received signal

$$r'(t) = r(t)e^{j(2\pi\delta ft+\theta)} \tag{5.19}$$

With (5.9) we obtain

$$
\begin{aligned}
y_{i,k} &= \frac{1}{T_{FFT}} \int_{t=kT}^{kT+T_{FFT}} r(t)e^{j(2\pi\delta ft+\theta)} e^{-j2\pi i(t-kT)/T_{FFT}} \, dt = \\
&\quad e^{j2\pi\theta} \frac{1}{T_{FFT}} \int_{t=kT}^{kT+T_{FFT}} \left[\int_{\tau=0}^{\tau_{max}} b(\tau)s(t-\tau)d\tau + n(t) \right] e^{j2\pi\delta ft} e^{-j2\pi i(t-kT)/T_{FFT}} \, dt
\end{aligned}
\tag{5.20}
$$

Repeating the derivation leading to (5.13), the received constellation points become

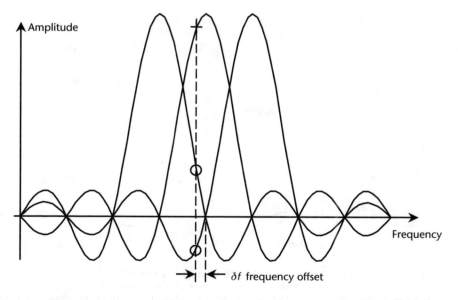

Figure 5.8 ICI arises in the case of a carrier synchronization error. This figure illustrates the spectra of three individual SCs. These spectra are superimposed in the OFDM signal spectrum.

$$y_{i,k} = e^{j(\theta + 2\pi\delta f k T)} \sum_{i'=-N/2}^{N/2-1} x_{i',k} h_{i',k} \frac{1}{T_{FFT}} \int_{u=0}^{T_{FFT}} e^{-j2\pi\left(\frac{i-i'}{T_{FFT}} - \delta f\right)u} du + n_{i,k} \qquad (5.21)$$

Due to the frequency error, the integral does not equal zero for $i \neq i'$; neither does it equal one for $i = i'$, as in the idealized case above; that is, the orthogonality between SCs has been partly lost. The evaluation of this expression yields two terms. The first (for $i = i'$) accounts for equal phase rotation and attenuation of all SCs; the second (for $i \neq i'$) describes the ICI.

$$y_{i,k} = e^{j(\theta + 2\pi\delta f k T)} x_{i,k} h_{i,k} \frac{1}{T_{FFT}} \int_{u=0}^{T_{FFT}} e^{j2\pi\delta f u} du +$$

$$e^{j(\theta + 2\pi\delta f k T)} \sum_{i'=-N/2}^{N/2-1} x_{i',k} h_{i',k} \frac{1}{T_{FFT}} \int_{u=0}^{T_{FFT}} e^{-j2\pi\left(\frac{i-i'}{T_{FFT}} - \delta f\right)u} du + n_{i,k} \qquad (5.22)$$

These expressions are valid for a frequency offset $\delta f < 0.5$ SC. For larger offsets, the transmitted data symbols $x_{i,k}$ would get shifted by one or more positions in the frequency direction; that is, the data symbol of the ith transmitted SC would appear at the $(i + \delta f_i)$-th SC at the receiver, where $\delta f_i = \text{round}(\delta f/F)$ is the integer part of the frequency error in SCs.

The ICI term can be seen as an additional noise term and can thus be represented as a degradation of SNR. Pollet et al. [10] have determined the amount of degradation for AWGN channels, and Moose [11] has done so for dispersive fading channels (see also [7]). Frequency offsets up to 2% of the SC spacing F are negligible, according to their results. Even 5% to 10% can be tolerated in many situations.

Evaluation of the phase rotation and attenuation due to a frequency error yields

$$y_{i,k} = x_{i,k} h_{i,k} \text{sinc}(\delta f T_{FFT}) \exp\left\{j\left[\theta + 2\pi\delta f(kT + T_{FFT}/2)\right]\right\} + n'_{i,k} \qquad (5.23)$$

using

$$\frac{1}{T_{FFT}} \int_{t=0}^{T_{FFT}} e^{j2\pi\delta f t} dt = \frac{1}{j2\pi\delta f T_{FFT}} \left[e^{j2\pi\delta f T_{FFT}} - 1\right]$$

$$= e^{j2\pi\delta f T_{FFT}} \frac{\sin\pi\delta f T_{FFT}}{\pi\delta f T_{FFT}} = e^{j\pi\delta f_{FFT}} \text{sinc } \delta f_{FFT} \qquad (5.24)$$

The noise term $n'_{i,k}$ includes the additional noise due to ICI.

Figure 5.9 depicts the rotation and distortion of the received signal constellation points for a carrier offset of $\delta f = F/16$, $\theta = 0$ and for QPSK modulation ("O"). The scattering of the resulting complex-valued signal constellations is caused by ICI. The figure also shows the projection of the continuous Fourier spectrum of one OFDM symbol on the complex plane, that is, the spectrum between the SC frequencies. This line results from the superposition of the continuous sinc spectra of individual SCs of one OFDM symbol. If a frequency offset is present, the DFT samples this

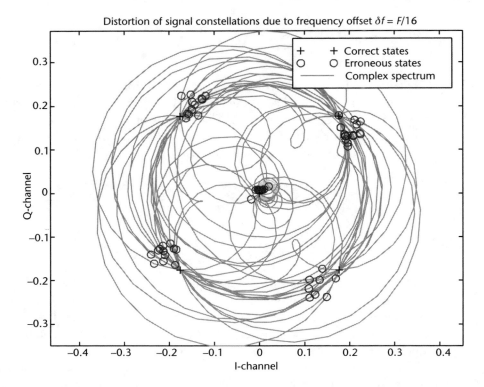

Figure 5.9 Phase rotation due to a carrier offset of 1/16 of the SC spacing. The received signal constellations distorted by ICI are shown.

spectrum at the wrong frequency instants, leading to ICI indicated in the figure by "O". Without frequency offset, the QPSK constellations are recovered perfectly, as seen from the points marked by "+".

5.2.4.3 Common Carrier and Timing Offset

Evaluating the above expressions for simultaneous timing (δt), frequency [δf, δf_i=round($\delta f/F$)] and phase (θ) offsets, the system model for the generalized case is obtained. It is written as

$$y_{i+\delta f_{i},k} = x_{i,k} h_{i,k} \operatorname{sinc}\!\left[(\delta f - \delta f_i F) T_{FFT}\right] e^{j\Psi_{i,k}} + n'_{i,k} \tag{5.25}$$

where the phase distortion due to synchronization errors is expressed by

$$\Psi_{i,k} = \theta + 2\pi\delta f\!\left(kT + \frac{T_{FFT}}{2} + \delta t\right) + 2\pi\delta t \frac{i}{T_{FFT}} \tag{5.26}$$

Note again that the noise variable $n'_{i,k}$ in (5.25) includes the noise caused by ICI, ISI, or both.

Often, the timing offset is expressed in samples, that is, $\delta t' = t/T_s$, and the frequency offset is normalized to the SC spacing $\delta f' = \delta f/F$. Using these symbols, the phase distortions are expressed by

$$\Psi_{i,k} = \theta + 2\pi\delta f'\left(\frac{1}{2} + k\frac{N + N_{guard} + N_{win}}{N} + \frac{\delta t'}{N}\right) + 2\pi\delta t'\frac{i}{N} \qquad (5.27)$$

5.3 Performance of an Uncoded OFDM System

In this section, analytical expressions are derived for the bit-error probabilities of uncoded OFDM systems over Rayleigh and Rician fading channels. The analysis is based on the work by Proakis ([1, Appendix B]). The obtained expressions are very general, allowing the evaluation of various modulation schemes and demodulation and channel-estimation techniques. The application of the formulas is shown for BPSK, QPSK, 8-PSK, and 16-QAM modulation, with coherent detection (perfect channel estimation) and differential detection.

The idealized system model derived in Section 5.2.3 is employed in this study. By incorporating into the system model SNR degradation due to synchronization errors, Doppler spread, or phase noise (which cause ICI, ISI, or both), or by incorporating the MSE of channel-estimation techniques, the effect of these impairments on the BER can be analyzed. The systematic phase rotations induced by synchronization errors must be considered as well. While we leave the evaluation of SNR degradations for future work, we briefly investigate the effects of phase rotations in the presence of (small) synchronization offsets.

In OFDM, differential detection can be employed in the time and frequency directions. From the BER of differential detection, it is evident that the time direction is preferable for the OFDM system parameters under investigation because the channel variations versus frequency are larger.

Section 5.3.1 reviews the OFDM system model and channel model. The derivation of the average BER is explained in Section 5.3.2. Performance results are given in Section 5.3.3.

5.3.1 Mathematical Modeling

The OFDM system models derived in Section 5.2 are used in this analysis. For the sake of simplicity we slightly change the indexing, however, as we only investigate the time or frequency direction at a given time. The system model of (5.14) becomes

$$y_k = x_k h_k + n_k \qquad (5.28)$$

where $\{x_k\}$ and $\{y_k\}$ are the transmitted and received signal constellation points (modulated data symbols), respectively, $\{h_k\}$ accounts for the correlated, complex-valued attenuation factors introduced by the time- and/or frequency-selective radio channel, and $\{n_k\}$ denotes samples of an AWGN process with $E\{|n_k|^2\} = \sigma_N^2$. The index k can be used as a time or frequency index, depending on the system aspect under investigation. The attenuation factors thereby constitute the time or frequency TF of the channel, respectively:

$$h_k = \begin{cases} H(f, kT) & \text{at given } f \\ H(kF, t) & \text{at given } t \end{cases} \tag{5.29}$$

where T is the duration of an OFDM symbol, including the guard and windowing intervals, and F denotes the frequency spacing between adjacent OFDM SCs.

The channel model is introduced into the analysis by considering respectively the spaced-time and spaced-frequency correlation functions of the (WSSUS; [1, 12, 13]) channel.

The FD channel model is used to describe the frequency variability. Thereby, we confine ourselves to the case of the exponentially decaying delay power spectrum, where a direct relation can be given between the channel parameters {P_0−average power, K−Rician factor, and τ_{rms}−RDS} and the channel correlation function

$$\phi_H(\Delta f) = E\{H^*(f)H(f + \Delta f)\} = \frac{P_0}{K+1}\left(K + \frac{1}{1 + j2\pi\Delta f \tau_{rms} K_1}\right) \tag{5.30}$$

In this equation, $K_1 = (K + 1)/\sqrt{2K + 1}$, Δf is the frequency lag, and * denotes the complex conjugate. The NRP (average power) is defined as $P_0 = E\{|h_k|^2\}$.

To model the time variability, the so-called Jakes Doppler spectrum can be used [14], augmented by an LOS component $\rho e^{j(2\pi f_\rho t + \theta_\rho)}$ at a given Doppler frequency f_ρ. Such a Doppler spectrum corresponds to a spaced-time correlation function

$$\phi_H(\Delta t) = E\{H^*(t)H(t + \Delta t)\} = \frac{P_0}{K+1}\left(Ke^{j2\pi f_\rho \Delta t} + J_0(2\pi f_m \Delta t)\right) \tag{5.31}$$

where $J_0(\bullet)$ denotes the zeroth-order Bessel function of the first kind, Δt is the time lag, and f_m is the maximum Doppler frequency. ($f_m = v_m/\lambda = v_m f_c/c$, where v_m is the mobile's velocity, λ is the wavelength, f_c is the carrier frequency, and c is the speed of light.)

5.3.2 Analytical Evaluation of the BER

Analytical expressions for the BER are derived in this section. Following [1], we start our analysis with defining the symbol transmitted as $x_{k,i}$, which is an element of the symbol set {$x_{k,m}$}, $m = \{1, 2, ..., M\}$. (M is the order of the modulation scheme.) At the receiver's site, an optimum detector will choose the symbol $x_{k,n} \in \{x_{k,m}\}$, which minimizes the distance metric

$$M_d(x_{k,n}) = \left|y_k - \hat{h}_k x_{k,n}\right|^2 \tag{5.32}$$

This symbol is assumed most likely to be the transmitted symbol. The term $\hat{h}_k x_{k,n}$ in this equation accounts for the channel estimation. An error occurs when the metric calculated for a symbol $x_{k,n} \neq x_{k,i}$ is smaller than the metric for the transmitted symbol $x_{k,i}$. The probability of this event is written as

$$P_e = \Pr\left\{ M_d\left(x_{k,n} \right) < M_d\left(x_{k,i} \right) \right\} = \Pr\{D < 0\} \tag{5.33}$$

where $D = M_d(x_{k,n}) - M_d(x_{k,i})$ is called the decision variable. Using (5.32), D becomes

$$D = y_k \hat{h}_k^* \left(x_{k,i}^* - x_{k,n}^* \right) + y_k^* \hat{h}_k \left(x_{k,i} - x_{k,n} \right) + \left| \hat{h}_k \right|^2 \left(\left| x_{k,n} \right|^2 - \left| x_{k,i} \right|^2 \right) \tag{5.34}$$

From the channel and system models, y_k is known to be a complex Gaussian RV. The same holds for $\hat{h}_k$, which is an estimate of the TF $H(f,t)$. Thus, the decision variable D is a special case of the generic quadratic form (see [1, Appendix B])

$$D = \sum_{l=1}^{L} \left(A|X_l|^2 + B|Y_l|^2 + CX_l Y_l^* + C^* X_l^* Y_l \right) \tag{5.35}$$

where X_l and Y_l are complex-valued Gaussian RVs, and A, B, and C are constants. In our case $L=1$, considering one transmitted symbol over one (sub) channel.[1]

The error probability is the probability that $D < 0$, which is evaluated in [1, Appendix B]. This probability is denoted as the integral over the PDF of D

$$P_e = \Pr\{D < 0\} = \int_{-\infty}^{0} p(D)dD \tag{5.36}$$

For $L=1$, the solution to this integral is written [1]

$$P_e = Q_1(a,b) - \frac{v_2/v_1}{1+v_2/v_1} I_0(ab) e^{-\frac{1}{2}(a^2+b^2)} \tag{5.37}$$

where $I_n(x)$ is the nth-order modified Bessel function of the first kind and $Q_1(a,b)$ is Marcum's Q function, which can be expressed in terms of Bessel functions as

$$Q_1(a,b) = e^{-\frac{1}{2}(a^2+b^2)} \sum_{n=0}^{\infty} (a/b)^n I_n(ab), \quad b > a > 0 \tag{5.38}$$

The parameters a, b, v_1, and v_2 must be related to the moments of X_l and Y_l and to the constants A, B, and C. As given in [1], these are obtained from

---

1.   The equation for $L>1$ allows for the evaluation of diversity schemes [1, 15–17].

$$a = \left[\frac{2v_1^2 v_2 \left(\alpha_1 v_2 - \alpha_2 \right)}{\left(v_1 + v_2 \right)^2} \right]^{1/2}$$

$$b = \left[\frac{2v_1 v_2^2 \left(\alpha_1 v_1 - \alpha_2 \right)}{\left(v_1 + v_2 \right)^2} \right]^{1/2}$$

$$v_{1,2} = \sqrt{w^2 + \frac{1}{4\left(\Psi_{xx}\Psi_{yy} - \left| \Psi_{xy} \right|^2 \right)\left(|C|^2 - AB \right)}} \mp w$$

(5.39)

$$w = \frac{A\Psi_{xx} + B\Psi_{yy} + C\Psi_{xy}^* + C\Psi_{xy}^*}{4\left(\Psi_{xx}\Psi_{yy} - \left| \Psi_{xy} \right|^2 \right)\left(|C|^2 - AB \right)}$$

$$\alpha_1 = 2\left(|C|^2 - AB \right)\left(\left| \overline{X}_1 \right|^2 \Psi_{yy} + \left| \overline{Y}_1 \right|^2 \Psi_{xx} - \overline{X}_1^* \overline{Y}_1 \Psi_{xy} - \overline{X}_1 \overline{Y}_1^* \Psi_{xy}^* \right)$$

$$\alpha_2 = A\left| \overline{X}_1 \right|^2 + B\left| \overline{Y}_1 \right|^2 + C\overline{X}_1^* \overline{Y}_1 + C^* \overline{X}_1 \overline{Y}_1^*$$

These equations are applied to our problem by comparing (5.34) and (5.35). Letting $Y_1 = y_k$ and $X_1 = \hat{h}_k$ in (5.34), the constants $A = \left| x_{k,n} \right|^2 - \left| x_{k,i} \right|^2$, $B = 0$, and $C = x_{k,n} - x_{k,i}$ are found, representing the properties of the modulation scheme. The behaviors of the channel and of the channel-estimation technique will be expressed by the first and second moments of the RVs X_1 and Y_1. These are

$$\overline{X}_1 = E\left\{ \hat{h}_k \right\}$$

$$\overline{Y}_1 = E\left\{ y_k \right\}$$

$$\Psi_{xx} = \frac{1}{2}\left[E\left\{ \left| \hat{h}_k \right|^2 \right\} - \left| \overline{X}_1 \right|^2 \right]$$

(5.40)

$$\Psi_{yy} = \frac{1}{2}\left[E\left\{ \left| y_k \right|^2 \right\} - \left| \overline{Y}_1 \right|^2 \right]$$

$$\Psi_{xy} = \frac{1}{2}\left[E\left\{ \hat{h}_k y_k^* \right\} - \overline{X}_1 \overline{Y}_1^* \right]$$

The derivation of these parameters from the channel and system definitions is given in the following section. Coherent and differential detection are investigated.

5.3.2.1   Application of the Mathematical Models

Coherent Detection with Perfect Channel Estimation
The kth symbol received is defined in (5.28) as $y_k = x_k h_k + n_k$. Perfect channel estimation means that the receiver has exact knowledge of the attenuation factor h_k, denoted by $\hat{h}_k = h_k$. Considering the transmitted symbol $x_{k,i}$ as a constant yields

$$\overline{X}_1 = E\{h_k\} = \rho \cdot e^{j\theta_\rho}$$

$$\overline{Y}_1 = E\{x_{k,i}h_k + n_k\} = x_{k,i}E\{h_k\} + E\{n_k\} = x_{k,i}\rho \cdot e^{j\theta_\rho}$$

$$\Psi_{xx} = \frac{1}{2}\left[E\{|h_k|^2\} - |\overline{X}_1|^2\right] = \frac{1}{2}\left[P_0 - \rho^2\right] \tag{5.41}$$

$$\Psi_{yy} = \frac{1}{2}\left[E\{|x_{k,i}h_k + n_k|^2\} - |\overline{Y}_1|^2\right] = \frac{1}{2}\left[|x_{k,i}|^2(P_0 - \rho^2) + \sigma_N^2\right]$$

$$\Psi_{xy} = \frac{1}{2}\left[E\{h_k(x_{k,i}h_k + n_k)^*\} - \overline{X}_1\overline{Y}_1^*\right] = \frac{1}{2}x_{k,i}^*[P_0 - \rho^2]$$

where $\rho e^{j\theta_\rho}$ is the LOS-component, with arbitrary phase θ_ρ and with an amplitude defined by $\rho^2 = P_0 K/(K+1)$.

Differential Detection
With differential detection, the decision for the received symbol y_k is made based on the adjacent symbol $y_{k-1} = x_{k-1}h_{k-1} + n_{k-1}$. For phase modulation schemes, this can be seen as a detection based on the channel estimate $\hat{h}_k = y_{k-1}/x_{k-1} = h_{k-1} + n_{k-1}/x_{k-1} = h_{k-1} + n'_k$, where $E\{|n'_k|^2\} = \sigma_{N'}^2$. Note that $\sigma_{N'}^2 = \sigma_N^2$, if the magnitude of x_{k-1} is one. The parameters $\overline{Y}_1$ and Ψ_{yy} are then equal, as in (5.41). The additional noise term n'_k, the correlation between h_k and h_{k-1}, and the Doppler shift of the LOS-component are expressed in

$$\overline{X}_1 = E\{h_{k-1} + n'_k\} = \rho \cdot e^{j(\theta_\rho - 2\pi f_\rho T)}$$

$$\Psi_{xx} = \frac{1}{2}\left[NRP + \sigma_{N'}^2 - \rho^2\right] \tag{5.42}$$

$$\Psi_{xy} = \frac{1}{2}x_{k,i}^*\left[E\{h_{k-1}h_k^*\} - \rho^2 e^{j2\pi f_\rho T}\right]$$

For evaluating differential detection in the frequency direction, let $T = 0$. Using the channel correlation functions given in Section 5.3.1, the correlation Ψ_{xy} between the attenuation factors at two adjacent symbols becomes

$$\Psi_{xy} = \frac{1}{2}x_{k,i}^*\frac{P_0}{K+1}\begin{cases}\dfrac{1}{1 - j2\pi F\tau_{rms}K_1} & \text{in frequency} \\[2mm] J_0(2\pi f_m T) & \text{in time}\end{cases} \tag{5.43}$$

Note that the influence of the channel variability is expressed by this correlation value only, which is defined by the parameter products $\tau_{rms}F$ and $f_m T$, for the two cases under consideration.

Performance results are given in Section 5.3.3. It will be shown that differential detection in the time direction is more robust than the FD variant for the OFDM system under consideration, that is, for a wideband indoor WLAN.

5.3.2.2 Application to Different Modulation Schemes

Assigning different constellation values to the variable $x_{k,n} \neq x_{k,i}$, the probability can be calculated that an erroneous symbol $x_{k,n}$ has been detected while the symbol $x_{k,i}$ was transmitted. This allows, for many modulation schemes, an exact calculation of the BER; for others, it allows the evaluation of close approximations [15–17].

In the following analysis we assume that all possible transmitted symbols $x_{k,i} \in \{x_{k,m}\}$ occur with equal probability.

BPSK and QPSK

Exact results can be obtained for BPSK and QPSK modulation. The signal constellations for these techniques are depicted in Figure 5.10. For both schemes it is sufficient to consider (any) one transmitted symbol, due to symmetries. This symbol will be the +1, taken from the set $\{x_{k,m}\} = \{1,-1\}$ for BPSK, and from $\{x_{k,m}\} = \{1, j, -1, -j\}$ for QPSK. Note that $|x_{k,m}|^2 = 1$ for both modulation types.

BPSK is analyzed by evaluating the parameters A and C for $x_{k,i} = 1$ and $x_{k,n} = -1$. The bit-error probability is equal to the symbol-error probability P_e [see (5.37)].

Gray-coded QPSK transmits two bits per symbol on orthogonal carriers (I- and Q-components). Thus, the error probabilities can be analyzed independently, and the BER equals their average. Suitable parameters for A and C are found, for example, by assigning $x_{k,i} = 1$ and $x_{k,n} = \{j, -j\}$.

Calculating the probability that $x_{k,n}$ has been detected, provided $x_{k,i}$ was transmitted, the I/Q-plane is divided in two parts. An error occurs when the received symbol falls within the half plane closer to $x_{k,n}$ than to $x_{k,i}$. No error occurs otherwise (see Figure 5.10). Note that in the case of QPSK, it is not necessary to evaluate the two-error event explicitly. The overlapping one-error events account for one error each in this region; thus, the two-error event is included automatically. This may seem

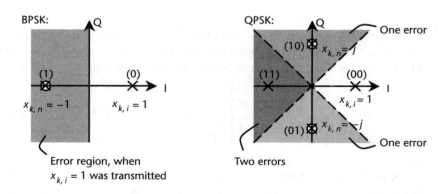

Figure 5.10 Selection of $x_{k,i}$ and $x_{k,n}$ for the performance evaluation of BPSK and QPSK.

like an advantage because computational complexity is reduced; however, when evaluating higher-order modulation schemes, many of those half-planes will overlap, and it is sometimes not possible to obtain the exact number of errors for all decision regions. This will be seen in the following case.

8-PSK

Upper and lower bounds on the BER can be calculated for 8-PSK. An exact calculation is not possible because the eight signal states are not separable in the two orthogonal dimensions of the I/Q-plane.

Due to symmetries it is again sufficient to consider one transmitted symbol, $x_{k,i} = 1$. Figure 5.11 illustrates how errors occur in estimating error probabilities. The signal constellations are shown in Figure 5.11(a), together with the exact numbers of errors for each decision region. (Errors are denoted by ε.) Figure 5.11(b) shows the actual numbers of errors for each of these regions, when three different error states $x_{k,n}$ are evaluated and averaged. Clearly, too few errors are considered in some of the decision regions, while too many are considered in others. Thus, the computational results are a (close) approximation. The most likely errors, however, are appropriately treated.

16-QAM

16-QAM can be evaluated without any error. This involves considering 4 different transmitted symbols occurring with equal probabilities and 24 error events. Some of them must be subtracted in order to account for overlapping decision regions. Table 5.1 lists a possible set of symbols $x_{k,i}$ and $x_{k,n}$ to be used. The complex signal constellations x_k are denoted $(\text{Re}\{x_k\},\text{Im}\{x_k\})$. Error events whose probability must be subtracted in the final result are written $(\text{Re}\{x_k\},\text{Im}\{x_k\})^{-1}$. All values must be divided by $\sqrt{10}$ to have an average power of one. Figure 5.12 illustrates the signal

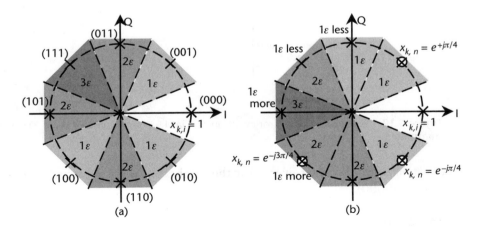

Figure 5.11   Error regions for 8-PSK when $x_{k,i} = 1$ was transmitted: (a) signal constellations and correct number of errors for each decision range, and (b) approximation by evaluating error probabilities from the three error states $x_{k,n}$ shown. In some error regions, one extra error is considered; in other regions, one error is missed (indicated as 1 more and 1 less)

Table 5.1 Transmitted Symbols and Error Events for the Evaluation of 16-QAM Modulation

Transmitted Symbol $x_{k,i}$	Error Symbols $x_{k,n}$
(−3,3)	(−1,3), (3,3), (7,3)$^{-1}$, (−3,1), (−3,−3), (−3,−7)$^{-1}$
(−1,3)	(−3,3), (1,3), (5,3), (−1,1), (−1,−3), (−1,−7)$^{-1}$
(−3,1)	(−1,1), (3,1), (7,1)$^{-1}$, (−3,3), (−3,−1), (−3,−5)
(−1,1)	(−3,1), (1,1), (5,1), (−1,3), (−1,−1), (−1,−5)

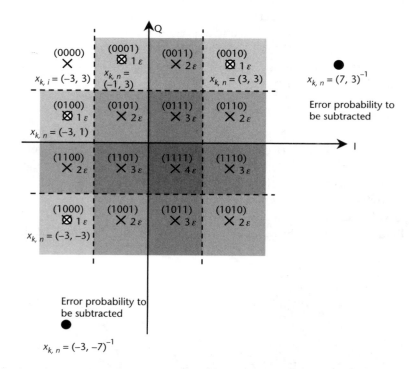

Figure 5.12 Illustration of the error events in 16-QAM when the symbol $x_{k,i}$=(3,3) was transmitted.

constellations and error events for the symbol $x_{k,i}$ = (−3,3), which carries the data symbol (0000).

16 Star-QAM

16 Star-QAM can be treated as a combination of 8-PSK and a binary amplitude modulation (AM). The binary AM is evaluated by transforming the I and Q variables to an $r^2 = I^2 + Q^2$ variable, resulting in similar expressions to the above defined ones. This is described in [16, 17].

5.3.3 Performance Results

Some observations can be made from the mathematical expressions derived above [see (5.41)–(5.43)]:

1. For coherent detection, the statistical parameters, and thus the performance results, only depend on P_0, ρ, and σ_N^2. In other words, the performance depends on the average SNR $SNR \propto P_0/\sigma_N^2$ and on the Rician K-factor $K = \rho^2/(P_0 - \rho^2)$.
2. The same holds in the limits $F \to 0$ or $T \to 0$ (i.e., for flat fading) for differential detection.
3. The performance of differential detection degrades for $F > 0$ (or $T > 0$) because of a systematic estimation error in $\hat{h}_k = h_{k-1} + n'_k$, because $h_{k-1} \neq h_k$. The parameter products $\tau_{rms} F$ and $f_m T$ define the degradation according to (5.43).

Performance results (average BER) for (1) and (2) and QPSK modulation are shown in Figure 5.13 as a function of the average SNR per bit (denoted E_b/N_0)[2] and as a function of K, where

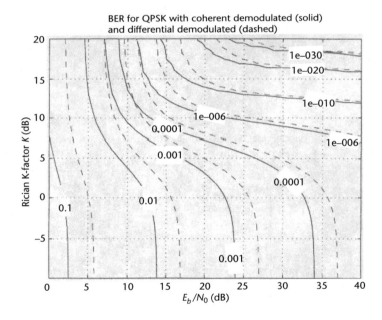

BER for QPSK with coherent demodulated (solid) and differential demodulated (dashed)

Figure 5.13 Performance of QPSK for coherent detection (perfect channel estimation) ("——") and for differential detection with $F = 0$, that is, with perfect correlation between adjacent SCs (flat fading) ("– – –").

2. Several SNR parameters are used in this chapter: The SNR denoted E_b/N_0 is the average SNR per data bit. It thus depends on the order M of the modulation scheme. The average SNR of the SC symbols, being independent of the modulation scheme, is written as SNR_{SC}. The SNR of the TD OFDM signal is written as SNR. This value is different from the previous ones because not all FFT points are used for data SCs.

$$\frac{E_b}{N_0} = \frac{E\left\{\left|x_{k,m}\right|^2\right\}}{\log_2(M)} \frac{P_0}{\sigma_N^2} \tag{5.44}$$

It is observed from the figure that the SNR required to achieve a certain BER-performance is dramatically increased for small K-factors (for Rayleigh channels). A 3-dB disadvantage of SNR is evident for the differential detection method because two noise processes with equal variance are present, the noise of the channel estimate and the noise of the data symbol to be detected.

Figure 5.14 presents a performance comparison of different modulation schemes. Note that the result for 16-QAM (differential) is a theoretical one because

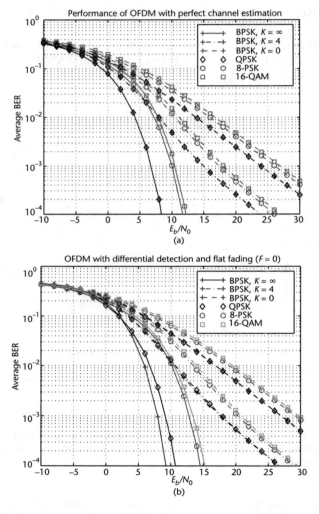

Figure 5.14 Performance of different modulation schemes: (a) coherent detection with perfect channel estimation, and (b) differential detection with $F = 0$, that is, perfect correlation between adjacent SCs (flat fading).

differential demodulation for this scheme is hard to accomplish. Again the advantage of a high K-factor is seen. With coherent detection, equivalent performance is obtained for BPSK and QPSK. This is not the case for differential detection on AWGN or Rician channels, where BPSK has an additional advantage of 1 to 2 dB over QPSK (see also [1]). It is important to note that twice the symbol energy is used with QPSK because two bits are transmitted per symbol. The higher-order modulation schemes (8-PSK and 16-QAM) require approximately 3 to 4 dB more signal power than QPSK.

Taking into account the channel variability, irreducible error floors arise (see Figure 5.15). Both versions of differential detection have been evaluated for Rayleigh fading channels, QPSK modulation, and for the following parameters. For detection in the frequency direction, the channel's RDS τ_{rms} is assumed to be three samples, which corresponds to a maximum delay spread of about 30 samples, assuming an exponentially decaying channel delay profile. For 128 FFT points, this value corresponds to about one quarter of the FFT time, which is also about the time duration that would be selected for the GI. It is seen that the irreducible error floor associated with such realistic parameters ($\tau_{rms}F=3/128$) lies around 10^{-2} (curve "O——O").

The time variability for differential detection in time direction corresponds to a mobile moving at 20 m/s, to a carrier frequency of 60 GHz, and to a symbol length of 1.3 μs. According to the system model, ICI due to Doppler spreading has been

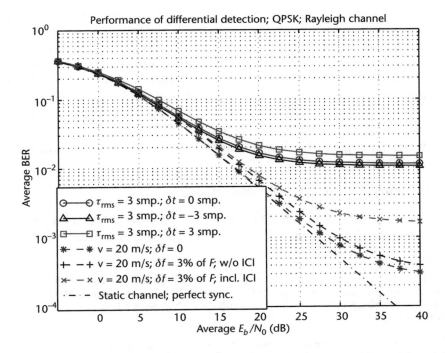

Figure 5.15 The channel variability leads to irreducible error floors for the differential modulation schemes. Differential QPSK is evaluated over Rayleigh fading channels.

neglected.[3] While the symbol duration assumed is rather short, the mobility considered is an order of magnitude higher than the expected mobility in an indoor WLAN system. Despite this, the error floor is much lower for this method of differential detection (curve "* – – *").

The other results depicted analyze the influence of synchronization errors. In the FD results, ICI and ISI have been neglected, which is exact as long as the channel IR remains within the GI. In other words, the performance impact results only from the systematic phase rotations that are—as the extended system model shows (Section 5.2.4)—due to the time-synchronization error. Such phase rotations mean for the differential detection techniques that the signal constellation points compared typically move closer together, which implies a performance degradation.

Note, however, that a negative timing offset equal to the channel's τ_{rms} slightly improves the performance because the channel also induces some systematic phase rotation, which is, in the case discussed, cancelled by the phase rotation due to timing offset.[4] The impact on the performance is small, however, for the offsets evaluated.

In the curves depicting the performance for the detection scheme in time direction, the impact of ICI due to a frequency offset is shown as well. ICI has been included using the approximation from [7]. It is seen that the impact of the phase distortion is evident in all SNR values, while ICI determines the error floor at high SNR.

In Figure 5.16, the performance of differential QPSK (in frequency direction) is shown as a function of E_b/N_0 and K, where τ_{rms} is a parameter. Since the maximum excess delay of the channel, which should not exceed the GI, is a function of τ_{rms} and K, all of these parameters are interrelated. The following definitions are introduced to get a set of general results. The FFT duration and GI duration are connected by a fixed factor, which is usually in the range of 4 to 10. The maximum excess delay of the channel can be written $\tau_{max} = 10\tau_{rms}K_1$. This leads to the normalized excess delay, defined as $T_m = \tau_{max}/T_{FFT} = 10\tau_{rms}K_1F$. In Figure 5.16, the performance of differential QPSK is shown for $T_m = \{0, 0.11, 0.27\}$. The curves for $T_m = 0$ allow a comparison with Figure 5.13. Especially for severely fading channels (low K-factors), the performance degradation is significant for the delay spreads considered.

5.4 Conclusions and Recommendations

The derivation of the OFDM system model has confirmed that data symbols can be transmitted independently over multipath fading radio channels. It has to be

---

3. That ICI is truly negligible for the system parameters selected is suggested from the comparison of the maximum Doppler frequency and the OFDM SC spacing. The former, being 4 kHz, is just 0.4% of the latter, which is 1 MHz. An approximate equation for the SNR degradation due to mobility can be found, for instance, in [7].

4. The progressive phase rotation due to a timing offset can be utilized for timing synchronization (see Section 6.2.7). Thereby, the systematic phase rotation due to the channel leads to a bias in the estimate. If this biased estimate is used for timing synchronization, optimum performance is achieved because the systematic phase rotations due to the channel and due to the bias compensate each other.

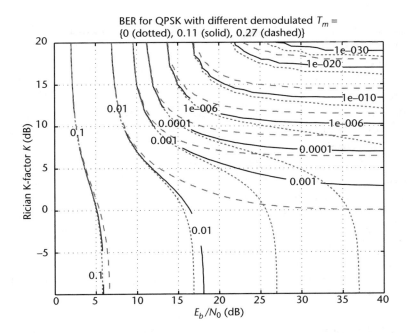

Figure 5.16 Performance of QPSK with differential detection in frequency direction. The maximum excess delay of the channel is related to the FFT time, expressed by T_m.

assumed, however, that the channel's maximum excess delay is shorter than the GI and that the system has been synchronized sufficiently. Small synchronization errors lead to systematic phase rotations of the data constellation points, a property that can be exploited for estimating synchronization offsets. If the timing- or frequency-synchronization error becomes too large, the orthogonality of the SCs is partly lost, and the SNR of the system is degraded; that is, ICI and ISI arise. ICI can also result from very fast channel variations (Doppler spreads) or from carrier phase jitters.

The system models presented can be utilized in analytical studies of various aspects of the OFDM technique, as, for instance, in the performance evaluation. The basic model introduced assumes perfect synchronization, while an extended model considers the phase rotations due to small synchronization offsets.

The performance analysis of an uncoded OFDM scheme is based on the classic formulas given by Proakis ([1, Appendix B]). Expressions are derived for the evaluation of different modulation schemes and for coherent and differential detection. The FD channel model for Rician fading channels has been applied. It allows performance results to be shown as a function of the channel parameters $\{P_0, K, \tau_{rms}\}$, the NRP, the Rician K-factor, and the RDS.

Assuming perfect channel estimation or, if differential schemes are applied, complete channel correlation, the performance is determined by P_0 and K. These parameters specify the average signal power and the depth of the fades. Better performance is thus achieved over channels having a higher K-factor because the fades are shallower.

Performing differential detection in the FD, a degradation of the results is seen due to the small differences of the channel TF at adjacent SCs (whose data symbols are compared). Since, for small frequency lags, there is a very strict relationship between this correlation function and the RDS, τ_{rms}, of the channel (in particular for Rayleigh fading channels), it is concluded that the performance degradation is well characterized by τ_{rms}. (To be exact, the performance is defined by the product $\tau_{rms}F$, where F is the SC spacing.) Imperfect timing synchronization also has an impact because systematic phase offsets are introduced between adjacent SCs.

For the low-mobility OFDM-based WLAN system under investigation, the correlation of subsequent symbols in time direction is much higher than the correlation of symbols on adjacent SCs. Therefore, it is recommended to apply differential detection in the time direction, not the frequency direction. In this case, systematic phase offsets are induced by imperfect carrier frequency synchronization.

By extending the OFDM system model, it becomes possible to analyze the imperfections of OFDM systems. Frequency synchronization offsets, for example, give rise to ICI, which can be accounted for by an additional noise term [11]. In a similar fashion, the impact of Doppler spreads, phase noise, or channel-estimation errors can be incorporated. The evaluation of such imperfections is a topic for future work. Using the original equations of [1], it is also possible to investigate diversity techniques (see, e.g., [15–17]).

References

[1]   Proakis, J. G., *Digital Communications*, 3rd ed., New York: McGraw-Hill, 1995.

[2]   Prasad, R., *Universal Personal Communications,* Norwood, MA: Artech House, 1998, Ch. 10.

[3]   van Nee, R., and R. Prasad, *OFDM for Wireless Multimedia Communications,* Norwood, MA: Artech House, 2000.

[4]   Edfors, O., et al., *An Introduction to Orthogonal Frequency-Division Multiplexing*, Research Report TULEA 1996:16, Division of Signal Processing, Luleå University of Technology, at http://www.sm.luth.se/csee/sp/publications.html.

[5]   Edfors, O., "Low-Complexity Algorithms in Digital Receivers," Ph.D. Thesis, Luleå University of Technology, September 1996.

[6]   Sandell, M., "Design and Analysis of Estimators for Multicarrier Modulation and Ultrasonic Imaging," Ph.D. Thesis, Luleå University of Technology, September 1996.

[7]   Speth, M., et al., "Optimum Receiver Design for Wireless Broadband Systems Using OFDM—Part I," *IEEE Trans. Communications*, Vol. 47, No. 11, November 1999, pp. 1668–1677.

[8]   Weinstein, S. B., and P. M. Ebert, "Data Transmission by Frequency-Division Multiplexing Using the Discrete Fourier Transform," *IEEE Trans. Communications Technology*, Vol. COM-19, No. 5, October 1971, pp. 628–634.

[9]   Peled, A., and A. Ruiz, "Frequency Domain Data Transmission Using Reduced Computational Complexity Algorithms," *Proc. IEEE Int. Conf. Acoust., Speech, Signal Processing*, Denver, CO, 1980, pp. 964–967.

[10] Pollet, P., M. van Bladel, and M. Moenclaey, "BER Sensitivity of OFDM Systems to Carrier Frequency Offset and Wiener Phase Noise," *IEEE Trans. on Communications,* Vol. 43, Nos. 2–4, February–April 1995, pp. 191–193.

[11] Moose, P. H., "A Technique for Orthogonal Frequency Division Multiplexing Frequency Offset Correction," *IEEE Trans. on Communications,* Vol. 42, No. 10, October 1994, pp. 2908–2914.

[12] Bello, P. A., "Characterization of Randomly Time-Variant Linear Channels," *IEEE Trans. on Commun. Systems,* Vol. CS-11, December 1963, pp. 360–393.

[13] Steele, R., *Mobile Radio Communications,* New York: John Wiley & Sons, 1992.

[14] Jakes, W. C., Jr., *Microwave Mobile Communications,* New York: John Wiley & Sons, 1974.

[15] Lu, J., et al., "BER Performance of OFDM-MDPSK Systems in Frequency-Selective Ricean Fading with Diversity Reception," *IEEE Trans. on Vehicular Technology,* Vol. 49, No. 4, July 2000, pp. 1216–1225.

[16] Dong, X., T. T. Tjhung, and F. Adachi, "Error Probability Analysis for 16 STAR-QAM in Frequency-Selective Ricean Fading with Diversity Reception," *IEEE Trans. on Vehicular Technology,* Vol. 47, No. 3, August 1998, pp. 924–935.

[17] Tjhung, T. T., et al., "BER Performance of 16 STAR-QAM in Ricean Fading with Diversity Reception," *Proc. PIMRC'94,* The Hague, the Netherlands, 1994, pp. 80–84.

CHAPTER 6

The Peak Power Problem

6.1 Introduction

An OFDM signal consists of a number of independently modulated SCs, which can give a large peak-to-average power (PAP) ratio when added up coherently. When N signals are added with the same phase, they produce a peak power that is N times the average power. This effect is illustrated in Figure 6.1. For this example, the peak power is 16 times the average value. The peak power is defined as the power of a sine wave with an amplitude equal to the maximum envelope value. Hence, an unmodulated carrier has a PAP ratio of 0 dB. An alternative measure of the envelope variation of a signal is the Crest factor, which is defined as the maximum signal value divided by the RMS signal value. For an unmodulated carrier, the Crest factor is 3 dB. This 3-dB difference between the PAP ratio and Crest factor also holds for other signals, provided that the center frequency is large in comparison with the signal bandwidth.

A large PAP ratio brings disadvantages like an increased complexity of the analog-to-digital (A/D) and digital-to-analog (D/A) converters and a reduced efficiency of the RF power amplifier. To reduce the PAP ratio, several techniques have

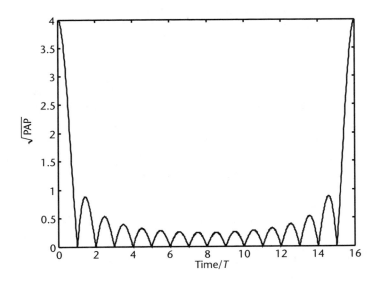

Figure 6.1 Square root of PAP ratio for a 16-channel OFDM signal, modulated with the same initial phase for all subchannels.

been proposed, which basically can be divided in three categories. First, there are signal distortion techniques, which reduce the peak amplitudes simply by nonlinearly distorting the OFDM signal at or around the peaks. Examples of distortion techniques are clipping, peak windowing, and peak cancellation. Second, there are coding techniques that use a special FEC code set that excludes OFDM symbols with a large PAP ratio. The third technique scrambles each OFDM symbol with different scrambling sequences and selecting the sequence that gives the smallest PAP ratio. This chapter discusses all of these techniques, but first analyzes the PAP ratio distribution function. This will give a better insight in the PAP problem and will explain why PAP reduction techniques can be quite effective.

6.2 Distribution of the PAP Ratio

For one OFDM symbol with N SCs, the complex baseband signal can be written as

$$x(t) = \frac{1}{\sqrt{N}} \sum_{n=1}^{N} a_n \exp(j\omega_n t) \tag{6.1}$$

Here, a_n are the modulating symbols. For QPSK, for instance, $a_n \in \{-1, 1, j, -j\}$. From the central limit theorem it follows that for large values of N, the real and imaginary values of $x(t)$ become Gaussian distributed, each with a mean of zero and a variance of 1/2. The amplitude of the OFDM signal therefore has a Rayleigh distribution, while the power distribution becomes a central chi-square distribution with two degrees of freedom and zero mean with a cumulative distribution given by

$$F(z) = 1 - e^{-z} \tag{6.2}$$

Figure 6.2 shows the probability that the PAP ratio exceeds a certain value. We can see that the curves for various numbers of SCs are close to the Gaussian distribution shown in Figure 6.2(d) until the PAP value comes within a few decibels of the maximum PAP level of 10log$N$, where N is the number of SCs.

Now, we want to derive the CDF for the peak power per OFDM symbol. Assuming the samples are mutually uncorrelated—which is true for nonoversampling—the probability that the PAP ratio is below some threshold level can be written as

$$P(PAPR \leq z) = F(z)^N = (1 - \exp(-z))^N \tag{6.3}$$

This theoretical derivation is plotted against simulated values in Figure 6.3 for different values of N.

The assumption made in deriving (6.3) that the samples should be mutually uncorrelated no longer holds when oversampling is applied. Because it seems quite difficult to come up with an exact solution for the peak power distribution, we propose an approximation by assuming that the distribution for αN SCs and oversampling can be approximated by the distribution for αN SCs without oversampling, with

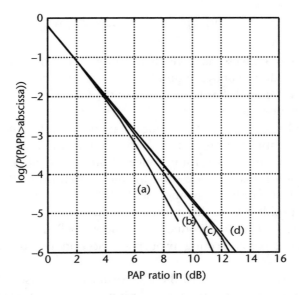

Figure 6.2   PAP distribution of an OFDM signal with (a) 12, (b) 24, (c) 48, and (d) an infinite number of SCs (pure Gaussian noise). Four times oversampling used in simulation; total number of simulated samples is 12 million.

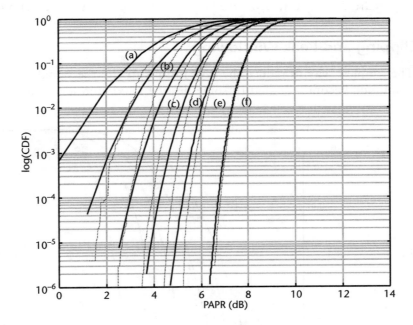

Figure 6.3   PAP distribution without oversampling for a number of SCs of (a) 16, (b) 32, (c) 64, (d) 128, (e) 256, and (f) 1,024. Dotted lines are simulated.

$\alpha > 1$. Hence, the effect of oversampling is approximated by adding a certain number of extra independent samples. The distribution of the PAP ratio is then given by

$$P(PAPR \leq z) = (1 - \exp(-z))^{\alpha - N} \tag{6.4}$$

Figure 6.4 gives the PAP distribution for different amounts of carriers for $\alpha = 2.8$. The dotted lines are simulated curves. We see in Figure 6.4 that (6.4) is quite accurate for $N > 64$. For large values of the CDF close to one (>0.5), however, (6.3) is actually more accurate.

From Figure 6.4, we can deduce that coding techniques to reduce the PAP ratio may be a viable option as reasonable coding rates are possible for a PAP ratio around 4 dB. For 64 SCs, for instance, about 10^{-6} of all possible QPSK symbols have a PAP ratio of less than 4.2 dB. This means that only 20 out of a total of 128 bits would be lost if only the symbols with a low PAP ratio were transmitted. However, the main problem with this approach is finding a coding scheme with a reasonable coding rate ($\geq 1/2$) that produces only these low PAP ratio symbols and that also has reasonable error-correcting properties. Section 6.5 describes a solution to this problem.

A different approach to the PAP problem is to use the fact that because large PAP ratios occur only infrequently, it is possible to remove these peaks at the cost of a slight amount of self-interference. Now, the challenge is to keep the spectral pollution of this self-interference as small as possible. Clipping is one example of a PAP reduction technique creating self-interference. In the next sections, two other techniques are described which have better spectral properties than clipping.

6.3 Clipping and Peak Windowing

The simplest way to reduce the PAP ratio is to clip the signal, such that the peak amplitude becomes limited to some desired maximum level. Although clipping is

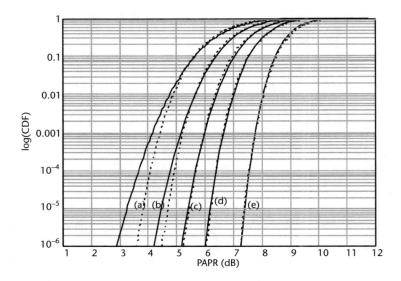

Figure 6.4 CDF of the PAP ratio (PAPR) for a number of SCs of (a) 32, (b) 64, (c) 128, (d) 256, and (e) 1,024. Solid lines are calculated; dotted lines are simulated.

definitely the simplest solution, there are a few problems associated with it. First, by distorting the OFDM signal amplitude, a kind of self-interference is introduced that degrades the BER. Second, the nonlinear distortion of the OFDM signal significantly increases the level of the out-of-band radiation. The latter effect can be understood easily by viewing the clipping operation as a multiplication of the OFDM signal by a rectangular window function that equals one if the OFDM amplitude is below a threshold and less than one if the amplitude needs to be clipped. The spectrum of the clipped OFDM signal is found as the input OFDM spectrum convolved with the spectrum of the window function. The out-of-band spectral properties are mainly determined by the wider spectrum of the two, which is the spectrum of the rectangular window function. This spectrum has a very slow roll off that is inversely proportional to the frequency.

To remedy the out-of-band problem of clipping, a different approach is to multiply large signal peaks with a certain nonrectangular window. In [1], a Gaussian-shaped window is proposed for this, but in fact any window can be used, provided it has good spectral properties. To minimize the out-of-band interference, ideally the window should be as narrowband as possible. On the other hand, the window should not be too long in the TD because that implies that many signal samples are affected, which increases the BER. Examples of suitable window functions are the cosine, Kaiser, and Hamming windows. Figure 6.5 gives an example of reducing the large peaks in OFDM with the use of windowing.

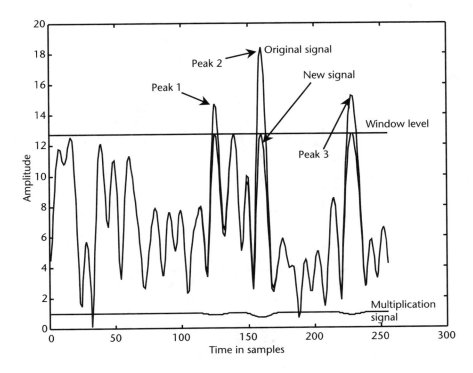

Figure 6.5 Windowing an OFDM time signal.

In Figure 6.6, the difference between clipping the signal and windowing the signal can be seen. Figure 6.7 shows how increasing the window width can decrease the spectral distortion.

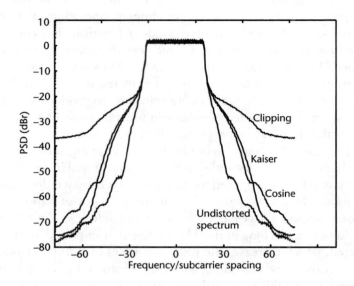

Figure 6.6 Frequency spectrum of an OFDM signal with 32 SCs with clipping and peak windowing at a threshold level of 3 dB above the RMS amplitude.

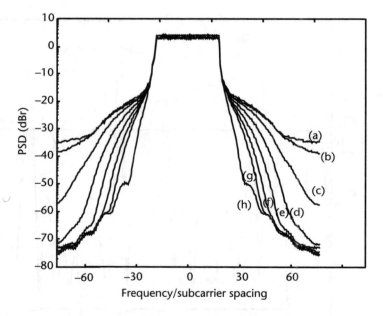

Figure 6.7 Frequency spectrum of an OFDM signal with 32 SCs with peak windowing at a threshold level of 3 dB above the RMS amplitude. Symbol length is 128 samples (4 times oversampled) and window length is (a) 3, (b) 5, (c) 7, (d) 9, (e) 11, (f) 13, and (g) 15 samples. Curve (h) represents the ideal OFDM spectrum.

Figure 6.8 shows packet-error ratio (PER) curves with and without clipping, using a rate 1/2 convolutional code with constraint length 7. The simulated OFDM signal uses 48 SCs with 16-QAM. The plots demonstrate that nonlinear distortion has only a minor effect on the PER; the loss in SNR is about 0.25 dB when the PAP ratio is decreased to 6 dB. When peak windowing is applied, the results are slightly worse (see Figure 6.9) because peak windowing distorts a larger part of the signal than clipping for the same PAP ratio.

6.3.1   Required Backoff with a Nonideal Power Amplifier

The previous section demonstrated that peak windowing is very effective in reducing the PAP ratio. This does not immediately tell us, however, what backoff is required for a practical power amplifier to attain an acceptable level of out-of-band radiation. The backoff is defined here as the ratio of the output power and the maximum output power (saturation power) with a sinusoidal input signal. Another definition frequently used in the literature uses the power at the 1-dB compression point instead of the saturation power. Because the 1-dB compression point is typically 1 to 3 dB lower than the maximum power level, depending on the amplifier TF, the backoff values according to the latter definition are 1 to 3 dB smaller than the values mentioned in this section.

To simulate a power amplifier, the following model is used for the AM/AM conversion [2]:

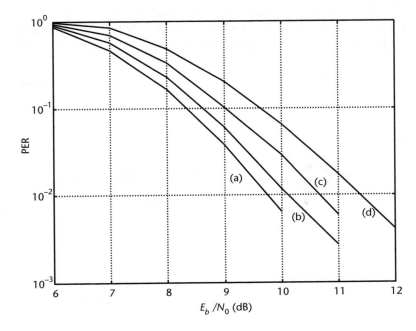

Figure 6.8   PER versus E_b/N_0 for 64-byte packets in AWGN. OFDM signal is clipped to a PAP ratio of (a) 16 (no distortion), (b) 6, (c) 5, and (d) 4 dB.

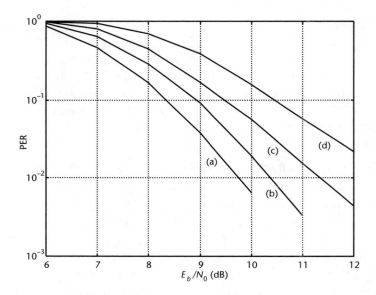

Figure 6.9 PER versus E_b/N_0 for 64-byte packets in AWGN. Peak windowing is applied with a window width of 1/16 of the FFT duration. The PAP ratio is reduced to (a) 16 (no distortion), (b) 6, (c) 5, and (d) 4 dB.

$$g(A) = \frac{A}{\left(1 + A^{2p}\right)^{\frac{1}{2p}}} \tag{6.5}$$

The AM/PM conversion of a solid-state power amplifier is small enough to be neglected. Figure 6.10 gives some examples of the TF for various values of p. A good approximation of existing amplifiers is obtained by choosing p in the range of

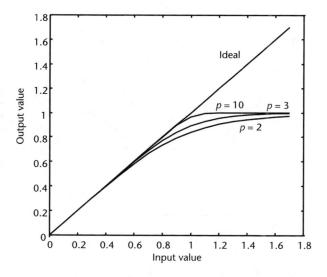

Figure 6.10 Rapp's model of AM/AM conversion.

2 to 3 [2]. For large values of p, the model converges to a clipping amplifier that is perfectly linear until it reaches its maximum output level.

Figure 6.11 shows the output spectra of an undistorted OFDM signal and the spectra of two distorted signals, assuming a highly linear amplifier model [$p = 10$ in (6.5)]. The backoff relative to the maximum output power was determined such that any significant distortion of the spectrum is at least 50 dB below the in-band spectral density. In this case, peak windowing gives a gain of almost 3 dB in the required backoff relative to clipping. This difference in backoff is much smaller than the difference in the PAP ratio at the input of the power amplifier. Without peak windowing, the PAP ratio is about 18 dB for the OFDM signal with 64 SCs. With peak windowing, it is reduced to approximately 5 dB. Hence, for the latter case, it is clear that the backoff of a highly linear amplifier must be slightly above this 5 dB to achieve a minimal spectral distortion. It is not true, however, that without peak windowing the backoff must be in the order of 18 dB for the same amount of distortion as with peak windowing. Because there is little energy in the signal parts that have a relatively large PAP ratio, distortion in those parts does not affect the spectrum that much. After peak windowing or any other PAP reduction technique, however, a significant part of the signal samples are close to the maximum PAP ratio (e.g., 5 dB). In this case, any distortion that is a decibel or so below this maximum produces more spectral distortion than clipping the original OFDM signal at 10 dB below its maximum PAP level, simply because for the latter, a much smaller fraction of the

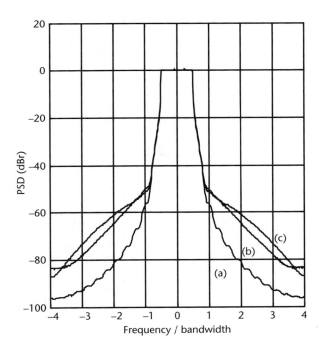

Figure 6.11 (a) Ideal OFDM spectrum for 64 SCs, (b) spectrum after highly linear amplifier (Rapp's parameter $p = 10$) with 8.7-dB backoff, and (c) spectrum using peak windowing with 5.9-dB backoff.

signal is affected. Thus, the more the PAP ratio is reduced by PAP-reduction techniques, the less tolerant the signal becomes to nonlinearities in the area of its maximum PAP ratio.

Figure 6.12 shows OFDM spectra for a more realistic amplifier model with $p = 3$. The target for undesired spectrum distortion has now been set to a less stringent level of 30 dB below the in-band density. The difference in backoff with and without peak windowing is now reduced to 1 dB. This demonstrates that the more spectral pollution can be tolerated, the less gain is achieved with PAP reduction techniques.

Figure 6.13 shows similar plots as Figure 6.12, but now for 256 SCs. This demonstrates that the required backoff with or without peak windowing is almost independent of the number of SCs, as long as this number is large compared with 1. In fact, the difference in backoff with and without peak windowing reduces slightly to 0.8 dB by going from 64 to 256 SCs.

6.3.2 Coding and Scrambling

A disadvantage of distortion techniques is that symbols with a large PAP ratio suffer more degradation, so they are more vulnerable to errors. To reduce this effect, FEC

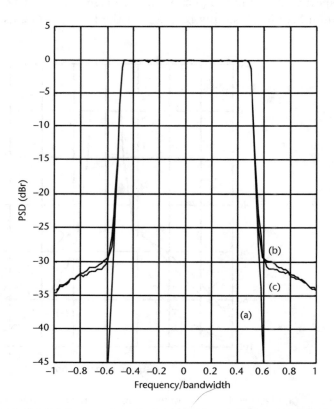

Figure 6.12 (a) Ideal OFDM spectrum for 64 SCs, (b) plain OFDM with 6.3-dB backoff and Rapp's parameter $p = 3$, and (c) peak windowing with 5.3-dB backoff.

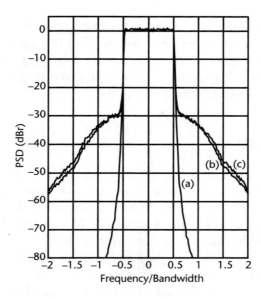

Figure 6.13 (a) Ideal OFDM spectrum for 256 SCs, (b) plain OFDM with 6.3-dB backoff and Rapp's parameter $p = 3$, and (c) peak windowing with 5.5-dB backoff.

coding can be applied across several OFDM symbols. By doing so, errors caused by symbols with a large degradation can be corrected by the surrounding symbols. In a COFDM system, the error probability is no longer dependent on the power of individual symbols, but rather on the power of a number of consecutive symbols. For example, assume that the FEC code produces an error if more than 4 out of every 10 symbols have a PAP ratio exceeding 10 dB.[1] Further, assume that the probability of having a PAP ratio larger than 10 dB is 10^{-3}. Then, the error probability of the peak cancellation technique is $1 - \sum_{i=0}^{3} \binom{10}{1} \left(10^{-3}\right)^{i} \left(1 - 10^{-3}\right)^{10-i} \cong 2 \cdot 10^{-10}$, which is much less than the 10^{-3} in the case where no FEC coding is used.

Although such a low symbol-error probability may be good enough for real-time circuit-switched traffic, such as voice, it may still cause problems for packet data. A packet with too many large PAP ratio symbols will have a large probability of error. Such packets occur only very infrequently, as shown above, but when they occur, they may never come through because every retransmission of the packet has the same large error probability. To solve this problem, standard scrambling techniques can ensure that the transmitted data between initial transmission and retransmissions are uncorrelated. To achieve this, the scrambler has to use a different seed for every transmission, which can be realized for instance by simply adding one to the seed after every transmission. Furthermore, the length of the scrambling sequence has to be in the order of the number of bits per OFDM symbol to

---

1. The simplifying assumption is made here that four symbols with reduced power always result in an error, while in reality there is always a certain error probability of less than one, depending on the SNR.

guarantee uncorrelated PAP ratios for different seeds. Different scrambling in every transmission will then guarantee independent PAP ratios for the OFDM symbols in retransmissions and, hence, independent error probabilities. For example, if the probability of a worst-case packet is 10^{-6}, the probability that it will not come through within two transmissions is 10^{-12}.

6.4 Peak Cancellation

The key element of all distortion techniques is to reduce the amplitude of samples whose power exceeds a certain threshold. In the case of clipping and peak windowing, this was done by a nonlinear distortion of the OFDM signal, which resulted in a certain amount of out-of-band radiation. This undesirable effect can be avoided by performing a linear peak-cancellation technique, whereby a time-shifted and scaled-reference function is subtracted from the signal, such that each subtracted reference function reduces the peak power of at least one signal sample. By selecting an appropriate reference function with approximately the same bandwidth as the transmitted signal, it can be assured that the peak power reduction will not cause any out-of-band interference. One example of a suitable reference signal is a sinc function. A disadvantage of a sinc function is that it has infinite support. Hence, for practical use, it has to be time limited in some way. One way to do this without creating unnecessary out-of-band interference is to multiply it by a windowing function, for instance, a raised cosine window. Figure 6.14 shows an example of a reference function, obtained by multiplication of a sinc function and a raised cosine

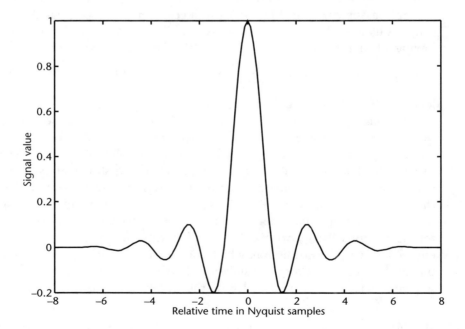

Figure 6.14   Sinc reference function, windowed with a raised cosine window.

window. If the windowing function is the same as that used for the windowing of the OFDM symbols, then it is assured that the reference function has the same bandwidth as the regular OFDM signals. Hence, peak cancellation will not degrade the out-of-band spectrum properties. By making the reference signal window narrower, a trade-off can be made between less complexity in the peak-cancellation calculations and some increase in the out-of-band power. The peak-cancellation method was first published in [3], and later was independently described in [4].

Peak cancellation can be performed digitally after generation of the digital OFDM symbols. It involves a peak power (or peak amplitude) detector, a comparator to see if the peak power exceeds some threshold, and a scaling of the peak and surrounding samples. Figure 6.15 shows the block diagram of an OFDM transmitter with peak cancellation. Incoming data is first coded and converted from a serial bit stream to blocks of N complex signal samples. On each of these blocks, an IFFT is performed. Then, a cyclic prefix is added, extending the symbol size to $N + N_G$ samples. After parallel-to-serial (P/S) conversion, the peak cancellation procedure is applied to reduce the PAP ratio. It is also possible to do peak cancellation immediately after the IFFT and before the cyclic prefix and windowing. Except for the peak cancellation block, there is further no difference from a standard OFDM transmitter. For the receiver, there is no difference at all, so any standard OFDM receiver can be used.

In the previous figures, the peak cancellation was done after P/S conversion of the signal. It is also possible to do the cancellation immediately after the IFFT, as depicted in Figure 6.16. In this case, the cancellation is done on a symbol-by-symbol basis. An efficient way to generate the cancellation signal without using a stored reference function is to use a lowpass filter in the FD. In Figure 6.16, for each OFDM symbol, the samples that exceed some predefined amplitude are detected. Then, for each signal peak, an impulse is generated whose phase is equal to the peak phase and whose amplitude is equal to the peak amplitude minus the desired maximum amplitude. The impulses are then lowpass filtered on a symbol-by-symbol basis. Lowpass filtering is achieved in the FD by taking the FFT, setting all outputs to zero whose frequencies exceed the frequency of the highest SC, and then transforming the signal back by an IFFT.

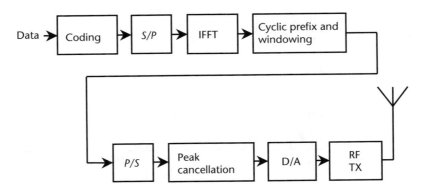

Figure 6.15   OFDM transmitter with peak cancellation.

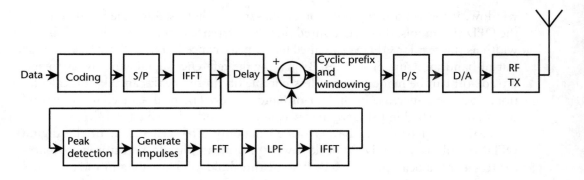

Figure 6.16   Peak cancellation using FFT/IFFT to generate cancellation signal.

Figure 6.17 shows an example of the cyclic reference function that is used in all methods that apply cancellation before adding the cyclic prefix and windowing. In fact, this reference signal itself is a valid OFDM signal, which is obtained in the case of an all-ones input to the IFFT.

Figure 6.18 shows an example of the signal envelopes of one arbitrary OFDM symbol and the corresponding cancellation signal. In this case, the cancellation signal actually consists of two separate sinc functions because one sinc function is not wide enough to reduce the peak. After subtraction, the peak amplitude is reduced to a maximum of 3 dB above the RMS value (see Figure 6.19).

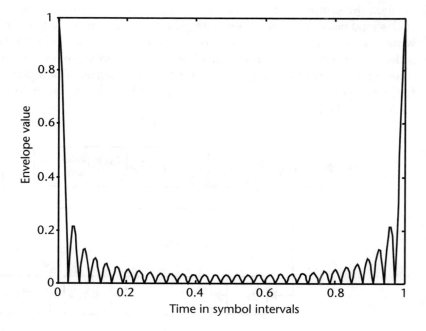

Figure 6.17   Envelope of cyclic reference function.

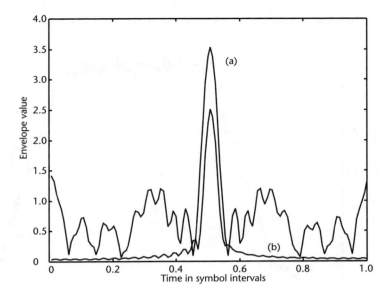

Figure 6.18 (a) OFDM symbol envelope, and (b) cancellation signal envelope.

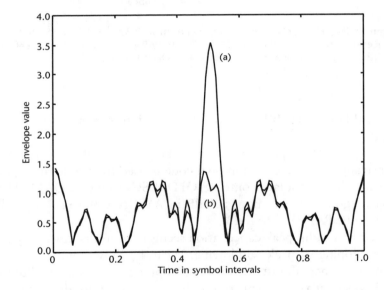

Figure 6.19 (a) OFDM symbol envelope, and (b) signal envelope after peak cancellation.

As an example of the peak cancellation technique, Figure 6.20 shows simulated PSDs for an OFDM system with 32 carriers. Without clipping or peak cancellation, the worst-case PAP ratio of this system is 15 dB, and the undistorted spectrum is depicted by curve (a). If the signal is clipped such that the PAP ratio reduces to 4 dB, a significant spectral distortion is visible; see curve (c). When peak cancellation is

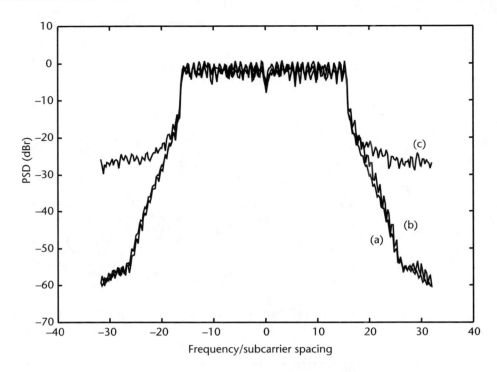

Figure 6.20 PSD for (a) undistorted spectrum with 32 SCs, PAP = 15 dB; (b) spectrum after peak cancellation to PAP = 4 dB; and (c) clipping to PAP = 4 dB. Reference cancellation function has a length equal to one quarter the length of an OFDM symbol.

applied [see curve (b)], a negligible distortion is present for the same PAP ratio of 4 dB.

Figure 6.21 depicts the effect of the peak cancellation on the PER. A rate 1/2, constraint length 7 convolutional code is used to encode the input bits. The coded bits are then modulated onto 48 OFDM SCs using 16-QAM. The curves show an SNR degradation of about 0.6 dB in AWGN when peak cancellation is used to reduce the PAP ratio to 6 dB.

At first sight, peak cancellation seems to be a fundamentally different approach than clipping or peak windowing. It can be shown, however, that peak cancellation is in fact almost identical to clipping followed by filtering. If a sampled OFDM signal $x(n)$ is clipped to reduce the PAP ratio, the output signal $r(n)$ can be written as

$$r(n) = x(n) - \sum_i a_i e^{j\varphi_i} \delta(n - \tau_i) \tag{6.6}$$

Here, a_i, φ_i, and τ_i are the amplitude, phase, and delay of the correction that is applied to the ith sample in order to reach the desired clipping level. Hence, it is possible to describe clipping as a linear process, even though it is not performed in that way in practice. Now, suppose the clipped signal is filtered by an ideal lowpass

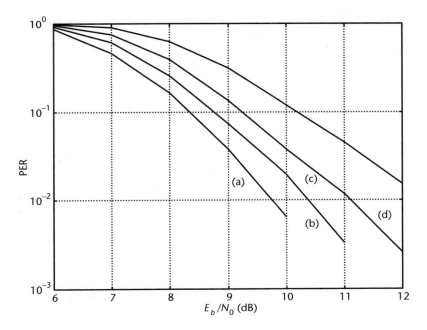

Figure 6.21   PER versus E_b/N_0 for 64-byte packets in AWGN. Peak cancellation is applied to reduce the PAP ratio to (a) 16 (no distortion), (b) 6, (c) 5, and (d) 4 dB.

filter with an IR of sinc($\pi n T$), where T is chosen such that the filter bandwidth is equal to or larger than the bandwidth of the OFDM signal. The filtered output is given by

$$r'(n) = x'(n) - \sum_i a_i e^{j\varphi_i'} \text{sinc}(\pi T(n - \tau_i)) \tag{6.7}$$

This expression is identical to a peak-cancellation operation, the only exception being that with peak cancellation, a sum of sinc functions is subtracted from the unfiltered OFDM signal $x(n)$, while in (6.7) we see a filtered signal $x'(n)$. In practice, however, also for peak cancellation, the OFDM signal needs to be filtered anyway to remove aliasing after the D/A conversion. Hence, for practical purposes, it may be concluded that peak cancellation has the same effect as clipping followed by filtering, which was proposed as a PAP reduction technique in [5].

As a final comparison of the three described signal distortion techniques, Figure 6.22 shows the PERs for an OFDM system with 48 SCs for which the PAP ratio is reduced to 5 dB. In addition to the three PAP reduction techniques, the non-linear amplifier model described in Section 6.3.1 was applied such that the output backoff of the transmitted OFDM signal was 6 dB. We can see from the figure that clipping (without filtering) performs slightly better than peak cancellation, and that peak windowing performs slightly worse than peak cancellation.

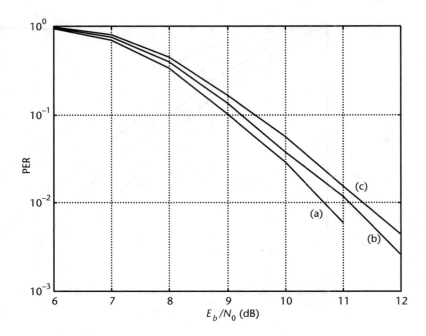

Figure 6.22 PER versus E_b/N_0 for 64-byte packets in AWGN. PAP ratio is reduced to 5 dB by (a) clipping, (b) peak cancellation, and (c) peak windowing.

6.5 PAP Reduction Codes

As Section 6.2 shows, only a small fraction of all possible OFDM symbols has a bad PAP ratio. This suggests another solution to the PAP problem, based on coding. The PAP ratio can be reduced by using a code that only produces OFDM symbols with PAP ratios below some desirable level. Of course, the smaller the desired PAP level, the smaller the achievable code rate is. Section 6.2, however, has already demonstrated that for a large number of SCs, a reasonable coding rate larger than 3/4 can be achieved for a PAP level of 4 dB. In [6], we find that for eight channels, a rate 3/4 code exists that provides a maximum PAP ratio of 3 dB. The results in [6] are based on an exhaustive search through all possible (QPSK) code words. Unfortunately, these results only tell us that a large number of code words exist; they do not tell us whether there is a structured way of encoding and decoding to generate a large part of these code words or what the minimum distance properties of the code are. However, [6] does mention the interesting fact that a large part of the codes found are Golay complementary sequences, which helps us define a structured way of generating PAP reduction codes. Golay complementary sequences are sequence pairs for which the sum of autocorrelation functions is zero for all delay shifts not equal to zero [7–9]. As [10] mentions, the correlation properties of complementary sequences translate into a relatively small PAP ratio of 3 dB when the codes are used to modulate an OFDM signal. Based on all of these hints toward Golay sequences, [11] presents a specific subset of Golay codes, together with decoding techniques that combine PAP reduction with good FEC capabilities. Based on this work, Golay codes have actually been implemented in a prototype 20-Mbps OFDM modem for the European Magic WAND project [12]. Fundamental studies on the coding properties of Golay sequences appeared in [13–16], proving code set sizes, distance properties, the relation to Reed-Muller codes, and many more interesting details.

A sequence x of length N is said to be complementary to another sequence y if the following condition holds on the sum of both autocorrelation functions:

$$\sum_{k=0}^{N-1}(x_k x_{k+i} + y_k y_{k+i}) \begin{aligned} &= 2N, \quad i = 0 \\ &= 0, \qquad i \neq 0 \end{aligned} \tag{6.8}$$

By taking the FTs of both sides of (6.8), the above condition is translated into

$$|X(f)|^2 + |Y(f)|^2 = 2N \tag{6.9}$$

Here, $|X(f)|^2$ is the power spectrum of x, which is the FT of its autocorrelation function. The DFT $X(f)$ is defined as

$$X(f) = \sum_{k=0}^{N-1} x_k e^{-j2\pi k f T_s} \tag{6.10}$$

Here, T_s is the sampling interval of the sequence x. From the spectral condition (6.9), it follows that the maximum value of the power spectrum is bounded by $2N$:

$$|X(f)|^2 \leq 2N \tag{6.11}$$

Because the average power of $X(f)$ (6.10) is equal to N, assuming that the power of the sequence x is equal to 1, the PAP ratio of $X(f)$ is bounded as

$$\text{PAP ratio} \leq \frac{2N}{N} = 2 \tag{6.12}$$

In an OFDM transmission, normally the IFFT is applied to the input sequence x. However, because the IFFT is equal to the conjugated FFT scaled by $1/N$, the conclusion that the PAP ratio is upper bounded by 2 is also valid when $X(f)$ is replaced by the inverse FT of the sequence x. Hence, by using a complementary code as input to generate an OFDM signal, the PAP ratio is guaranteed not to exceed 3 dB. Figure 6.23 shows a typical example of an OFDM signal envelope when using a complementary sequence. For this case of 16 channels, the PAP ratio is reduced by approximately 9 dB in comparison with the uncoded case of Figure 6.1.

6.5.1 Generating Complementary Codes

In [7–9], several coding rules are given for generating a set of complementary sequences, based on some starting complementary pair, the kernel. For complementary sequences of length 2, for instance, a possible kernel includes the pair 1,1 and

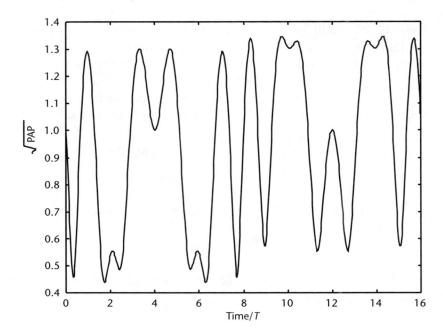

Figure 6.23 Square root of PAP ratio for a 16-channel OFDM signal, modulated with a complementary code.

1,−1. The basic coding rules for generating complementary codes from this kernel are as follows [7, 9]:

1. Interchanging both codes;
2. Reversing and conjugating second code;
3. Phase-rotating second code;
4. Phase-rotating elements of even order in both codes;
5. Phase-rotating first code;
6. Reversing and conjugating first code.

When rules 1 to 4 are applied, 16 different length 4 codes can be obtained for the case of four-phase modulation (see Table 6.1).

The number of codes can be extended to 64 by applying the rules 5 and 6, which gives the same result as applying four different phase shifts to the 16 codes. Hence, these four-symbol codes can easily be generated by using a 16-word-long lookup table to encode four bits, followed by a phase rotation to map a total of six bits onto all possible complementary codes.

Unfortunately, as indicated by the previous example, the six coding rules do not produce all complementary sequences unambiguously. This makes it difficult to find the size of the code set and to find a systematic way to produce complementary sequences. Thus, some other algorithm has to be found to generate complementary codes.

Table 6.1   Length 4 Complementary Codes

1	1	1	−1		1	1	j	−j
1	1	−1	1		1	1	−j	j
1	−1	1	1		1	−1	j	j
1	−1	−1	−1		1	−1	−j	−j
1	j	1	−j		1	j	j	1
1	j	−1	j		1	j	−j	−1
1	−j	1	j		1	−j	−j	1
1	−j	−1	−j		1	−j	j	−1

   Reference [9] shows that from one set of complementary sequences, others can be generated by multiplying the original sequences with columns of the DFT matrix. Although [9] only mentions this method to generate sets with longer code length by using the Kronecker product, it can also be used to generate different sequences with the same length by multiplying an original sequence elementwise with columns of the DFT matrix. It is easy to show that such multiplications do not change the correlation properties. Each DFT column is a delta function in the FD. Because multiplication in the TD is equivalent to a convolution in the FD, the power spectrum of a complementary sequence multiplied by a DFT column remains the same. Hence, its correlation function, which is the FT of the power spectrum, also remains the same, so that the outcome again is a complementary sequence.

   Another interesting remark in [9] is that complementary sequences can be multiplied by columns of the binary Walsh-Hadamard matrix without losing their complementary characteristics. Furthermore, [9] states that "if the code is an expansion of shorter lengths, an arbitrary phase angle can be added to all elements in any orthogonal subset." These operations turn out to be very useful in generating distinct codes.

   The coding algorithm for generating complementary sequences is now given by the following steps:

1. Make a kernel, that is, one complementary pair from which all other complementary sequences can be derived. For lengths equal to a power of two, kernels can easily be formed by using Golay's rule for length expansion. Starting with the length 2 sequence $A_1 B_1$, where $A_1 = 1$ and $B_1 = 1$, longer length codes can be formed by making $A_n B_n$ with $A_n = A_{n-1} B_{n-1}$ and $B_n = A_{n-1} -B_{n-1}$. In this way, codes of length 2^{n+1} are formed from the codes of length 2^n. For example, the following codes of up to length 16 can be obtained:

$$
\begin{aligned}
&\text{length 2: } A_1 B_1 = 1\ 1; \\
&\text{length 4: } A_2 B_2 = 1\ 1\ 1\ {-1}; \\
&\text{length 8: } A_3 B_3 = 1\ 1\ 1\ {-1}\ 1\ 1\ {-1}\ 1; \\
&\text{length 16: } A_4 B_4 = 1\ 1\ 1\ {-1}\ 1\ 1\ {-1}\ 1\ 1\ 1\ 1\ {-1}\ {-1}\ {-1}\ 1\ {-1};
\end{aligned}
\tag{6.13}
$$

2. Determine the number of orthogonal subsets. For length N codes, formed by the length expansion method described above, there are $\log_2 N$ orthogonal subsets, all of which can be given an arbitrary phase offset. The orthogonal subsets within a code are formed by all single elements, pairs, quads, and so forth, which are of even order. Thus, a length 16 code has four orthogonal subsets, consisting of all even elements, pairs, quads, and one octet. All of these can be given a different phase without changing the complementary characteristics of the code. Further, it is also possible to apply an arbitrary phase shift to the entire code. Hence, a complementary code set based on the kernel of (6.13) can be written as:

$$c = \left\{ \begin{array}{l} e^{j(\varphi_1+\varphi_2+\varphi_3+\varphi_4)}, e^{j(\varphi_1+\varphi_3+\varphi_4)}, e^{j(\varphi_1+\varphi_2+\varphi_4)}, -e^{j(\varphi_1+\varphi_4)}, \\ e^{j(\varphi_1+\varphi_2+\varphi_3)}, e^{j(\varphi_1+\varphi_3)}, -e^{j(\varphi_1+\varphi_2)}, e^{j\varphi_1} \end{array} \right\} \tag{6.14}$$

Notice that this code is actually implemented in a 20-Mbps OFDM modem for the Magic WAND project [12]. It is also used in the 11-Mbps IEEE 802.11 WLAN standard [17]. The latter is not an OFDM system, but here the benefit of using complementary sequences is their good aperiodic auto-correlation properties, which make it easier to build a receiver with sufficient robustness to multipath.

An alternative code description is to write the code phases as

$$\begin{bmatrix} \theta_1 \\ \theta_2 \\ \theta_3 \\ \theta_4 \\ \theta_5 \\ \theta_6 \\ \theta_7 \\ \theta_8 \end{bmatrix} = \begin{bmatrix} 1 & 1 & 1 & 1 & 0 \\ 1 & 0 & 1 & 1 & 0 \\ 1 & 1 & 0 & 1 & 0 \\ 1 & 0 & 0 & 1 & 1 \\ 1 & 1 & 1 & 0 & 0 \\ 1 & 0 & 1 & 0 & 0 \\ 1 & 1 & 0 & 0 & 1 \\ 1 & 0 & 0 & 0 & 0 \end{bmatrix} \begin{bmatrix} \varphi_1 \\ \varphi_2 \\ \varphi_3 \\ \varphi_4 \\ \pi \end{bmatrix} \tag{6.15}$$

The output code is given by $\exp(j \cdot 2\pi\theta_i/M)$, where θ_i is the coded phase and M is the size of the phase constellation. For BPSK ($M = 2$), the code set is equal to the Walsh-Hadamard codes, which is offset by the kernel, defined by the fourth column in (6.15).

3. Finally, a transformation can be applied that unfortunately cannot be described by simple multiplications or phase rotations. Instead, it can be described as an interleaving operation on the underlying shorter length codes used to make a longer length code [14]. For a length 8 sequence, for instance, two new length 8 codes can be generated by interleaving the first and second halves of the original code. Interleaving the code three times reproduces the original code. In general, a code with a length of 2^n can be interleaved $n - 1$ times before reproducing itself. The following shows three different codes produced by interleaving out of one length 8 code:

$$0:\ 1\ 1\ 1\ -1\ 1\ 1\ -1\ 1$$
$$1:\ 1\ 1\ 1\ 1\ 1\ -1\ -1\ 1 \tag{6.16}$$
$$2:\ 1\ 1\ 1\ -1\ 1\ -1\ 1\ 1$$

For a length 16 code, it turns out that except for four different codes that can be produced by interleaving the first and second halves of the code, more codes can be made by simultaneously interleaving the quarters of the code, giving a total of $3 \cdot 4 = 12$ different codes. The described coding rules can now be used to determine the size of complementary code sets. For an N-length code with M possible phases, the kernel can be multiplied by $1 + \log_2 N$ modified Walsh-Hadamard rows with M different phases. This gives a code set size of $M^{1+\log N}$. The amount of bits per code word can be expressed as $(1 + \log_2 N)\log_2 M$. For instance, a length 8 code with four possible phases gives eight bits per code word. The above numbers have not yet taken into account the interleaving rule, which adds another $\log_2([(\log_2 N)!]/2)$ bits to the total number of bits per symbol (for $N > 4$ with N being a power of 2). Notice that the interleaving rule does not necessarily produce an integer number of bits per encoded symbol.

6.5.2 Minimum Distance of Complementary Codes

In OFDM systems, the effects of multipath are mitigated by error-correction coding over the various subchannels. Thus, when using a PAP-reduction code, using this code for FEC as well would be very desirable. Otherwise, a separate code would be required, with the disadvantage of additional complexity and a reduction in the overall coding rate and spectral efficiency.

Therefore, the question arises as to what minimum distance the above-mentioned complementary sequences have. Looking at (6.15), we can state that if this is the only generating rule used, then $N/2 + 1$ correctly received symbols are always sufficient to calculate the $1 + \log_2 N$ phases used to generate the complementary sequence. This is because with $1 + N/2$ phase observations, it is always possible to form $1 + \log_2 N$ independent equations which can be used to solve for the $1 + \log_2 N$ unknown phases. In fact, there are a certain number of combinations of $1 + \log_2 N$ independent equations. The equations are independent only if each phase, except φ_1, is present in at least one and at most $\log_2 N$ equations. Since each phase, except φ_1, is present in exactly $N/2$ observations, $1 + N/2$ observations are sufficient to obtain at least one set of $1 + \log_2 N$ independent phase equations. Therefore, we can conclude that the minimum distance between two different complementary codes of length N is $N/2$ symbols, so it is possible to correct $N/4 - 1$ symbol errors or $N/2 - 1$ erasures.

The minimum Euclidean distance, which determines the performance in flat fading with additive noise, can be found by observing that a minimum distance between two code words is obtained if $N/2$ symbols have a minimum phase rotation of $2\pi/M$, where M is the number of phases. Thus, the minimum Euclidean distance d_{min} is

$$d_{min} = \sqrt{\frac{N}{2}} \left\| 1 - \exp\left(j\frac{2\pi}{M}\right) \right\| \tag{6.17}$$

For instance, for 8-PSK and eight channels, the minimum distance becomes 1.53, which is 6 dB larger than the distance of uncoded 8-PSK (=0.765). Because the rate of the length 8 complementary codes is 1/2, a maximum coding gain of 3 dB can be achieved compared with uncoded 8-PSK.

The above distance calculations are only valid for complementary codes generated without using the interleaving rule. Two codes formed by interleaving generally have a distance that is less than $N/2$ symbols. For $N = 8$, for instance, the interleaved codes (6.16) have a distance of only two symbols instead of four.

6.5.3 Maximum-Likelihood Decoding of Complementary Codes

This section describes an optimal decoding technique for specific subsets of complementary codes, based on generalized Walsh-Hadamard encoding. By generalized Walsh-Hadamard coding, we mean that for a length $N=2^n$ code, $n+1$ phases are encoded into 2^n output phases by adding the first phase to all code phases, the second to all odd code phases, the third to all odd pairs of code phases, and so on. For a length 8 code, for instance, the phase encoding is given by (6.15). For BPSK ($M = 2$), the coding reduces to normal Walsh-Hadamard coding. For this case, the efficient fast Walsh transform can be used to realize maximum-likelihood decoding. For larger constellation sizes, maximum-likelihood decoding seems less trivial. In the worst case, it would require M^{n+1} Euclidean distance calculations or correlations, giving a total number of operations of NM^{n+1} (complex multiplications and additions). There is, however, quite some redundancy in the calculation of all possible correlations, just as there are for the binary case. This means it is possible to reduce the complexity of the maximum-likelihood decoder by generalizing the fast Walsh transform to general phase constellations, as was first described in [18].

Figure 6.24 shows a butterfly that is used to calculate a 2-point binary fast Walsh transform. Using these butterflies, an N-point fast Walsh transform can be calculated with $N\log_2 N$ additions and subtractions.

Now consider a 4-PSK generalized Walsh-Hadamard code. For length 2, the transform can be depicted as a butterfly with two inputs and four outputs (see Figure 6.25).

The right side of the butterfly shows the sequences used to correlate with the input sequence.

Using this butterfly, a transform of double length N can be constructed by doing two transforms on half of the code length, plus an additional stage of 4^{n-1} butterflies. A length 4 transform, for instance, can be constructed as depicted in the Figure 6.26.

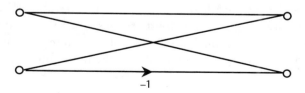

Figure 6.24 Butterfly of binary fast Walsh transform.

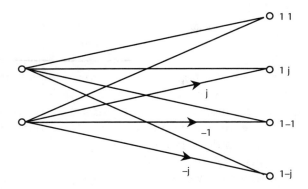

Figure 6.25 Butterfly of 4-PSK fast Walsh transform.

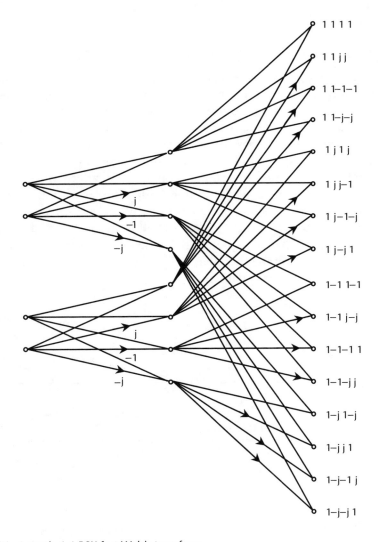

Figure 6.26 Length 4 4-PSK fast Walsh transform.

The 4 input points on the left are transformed into 16 output points by correlations with the complex sequences listed on the right.

The 4-point transform can be extended to an 8-point transform by performing two length 4 transforms on two groups of four samples, and adding an additional stage of 16 butterflies to produce the length 8 results. Figure 6.27 shows this 8-point transform, where the first two 4-point transforms are drawn as a line to simplify the picture. In reality, each 4-point transform has 16 outputs, which are combined with the 16 outputs of the other 4-point transform in four different ways. Hence, the total number of outputs is 64.

A length 8 transform needs 28 butterflies. Each butterfly requires 4 additions (the phase rotations are trivial for 4-PSK), so the total number of operations is 112 complex additions. The direct calculation method with 64 separate correlators requires 512 complex additions, so the fast transform reduces the complexity almost by a factor of five.

6.5.4 Suboptimal Decoding of Complementary Codes

For any coding technique to be successful, there must exist decoding techniques that are not too complex and with performance not too far from optimal maximum-likelihood decoding. For a complementary code for which the number of phases M is larger than two and the code length is larger than about eight, maximum-likelihood decoding quickly becomes too complex for practical implementation. Hence, we want to find suboptimal decoding techniques that are less complex to implement. One way to decode the phase that is applied to all alternating elements of a complementary code is to multiply the complex odd samples with the complex conjugate of the even samples. By summing the results, we obtain a vector that has the desired phase value. The same procedure can be followed for even and odd pairs, quads, and so on. The phase that is applied to the entire code has to be found by correcting the complex samples for all other phases. For the length 8 code with complex samples x_i, the phase equations are given by

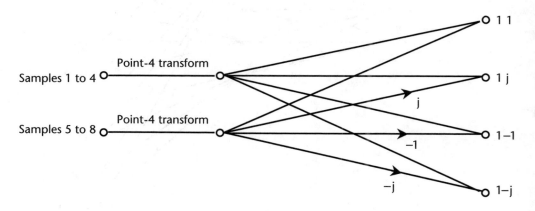

Figure 6.27 Length 8 4-PSK fast Walsh transform using length 4 transforms.

$$\varphi_2 = \arg\left\{ x_1 x_2^* + x_3 x_4^* + x_5 x_6^* + x_7 x_8^* \right\}$$

$$\varphi_3 = \arg\left\{ x_1 x_3^* + x_2 x_4^* + x_5 x_7^* + x_6 x_8^* \right\}$$

$$\varphi_4 = \arg\left\{ x_1 x_5^* + x_2 x_6^* + x_3 x_7^* + x_4 x_8^* \right\} \tag{6.18}$$

$$\varphi_1 = \arg\left\{ \begin{array}{l} x_1 e^{-j(\varphi_2+\varphi_3+\varphi_4)} + x_2 e^{-j(\varphi_3+\varphi_4)} + x_3 e^{-j(\varphi_2+\varphi_4)} + \\ x_4 e^{-j(\varphi_4)} + x_5 e^{-j(\varphi_2+\varphi_3)} + x_6 e^{-j(\varphi_3)} + x_7 e^{-j(\varphi_2)} + x_8 \end{array} \right\}$$

Here, arg{} means the calculation of the phase of a complex vector, and * denotes the complex conjugate. To convert the phases to bits, we have to make decisions for those constellation points that are closest to the phases found, just as we do in normal PSK.

There are some alternate ways to estimate the phase of the entire code word. In (6.18), the estimated phases were used to eliminate the phase rotations caused by all phases except for φ_1. The same effect can be achieved by multiplying the received code samples with complex conjugates of y_i, where y_i is the term within the arg{} expression of φ_i in (6.18):

$$\varphi_1 = \arg\left\{ \begin{array}{l} x_1 y_2^* y_3^* y_4^* + x_2 y_3^* y_4^* + x_3 y_2^* y_4^* + x_4 y_4^* + \\ x_5 y_2^* y_3^* + x_6 y_3^* + x_7 y_2^* + x_8 \end{array} \right\} \tag{6.19}$$

The disadvantage of this method is a certain noise enhancement due to the double and triple products of noisy phasors. A better estimate can be found by using only those terms that have no more than one phasor multiplication:

$$\varphi_1 = \arg\left\{ x_4 y_4^* + x_6 y_3^* + x_7 y_2^* + x_8 \right\} \tag{6.20}$$

Following the same argument, it is also possible to simplify (6.19) by using only terms with one or zero phase rotations:

$$\varphi_1 = \arg\left\{ x_4 e^{-j\varphi_4} + x_6 e^{-j\varphi_3} + x_7 e^{-j\varphi_2} + x_8 \right\} \tag{6.21}$$

The advantage of the above-described decoding technique is that it provides automatic weighting of the subchannels; erroneous channels with low amplitudes will only make a minor contribution to the phase estimates. In AWGN, the described technique performs 3 dB worse than optimal maximum-likelihood decoding, which can be argued as follows: The performance of maximum-likelihood decoding is determined by the minimum Euclidean distance, which is four times the distance of uncoded 8-PSK for the length 8 complementary code with 8-PSK. Looking at the decoding structure of (6.18), we can see that for each phase estimate, four or more vectors are added, which gives a 6-dB SNR improvement. Each of the added vectors, however, consists of a multiplication of two separate vectors with independent noise contributions. Hence, the detection SNR is only improved by 3 dB, as compared with 6 dB for a maximum-likelihood decoding technique. Note that the difference in maximum-likelihood decoding decreases in the case of frequency-selective channels. In the extreme case where four out of eight subchannels are completely lost, both will have the same symbol-error probability.

Except for the soft-decision technique described earlier, it is also possible to do hard-decision erasure decoding. In this case, four out of eight subchannels are erased, based on amplitude measurements obtained during training. Three subchannels can be erased arbitrarily; the fourth must be chosen such that all phase estimates in (6.18) have at least one element. Erasure decoding will fail if one of the unerased subchannels is in error. Thus, in AWGN, the bit-error probability is equal to that of uncoded 8-PSK, so there is a 6-dB loss compared with that of maximum-likelihood decoding. Again, this loss is less in the case of frequency-selective channels.

Figure 6.28 shows BER and PER for a single ATM cell packet versus mean E_b/N_0, averaged over a large (10^4) number of Rayleigh fading channels with an exponentially decaying PDP. The results clearly show that the combination of OFDM and complementary coding can efficiently exploit the frequency diversity of the channel for delay spreads of 10 ns or more. In this simulation, the use of 8-PSK length 8 complementary codes is assumed. Two independent codes together encode 24 bits into 16 OFDM channels. For a symbol duration of 1.2 μs, including a guard time of 400 ns, this gives a data rate of 20 Mbps. These parameters are used in the OFDM modem of Magic WAND.

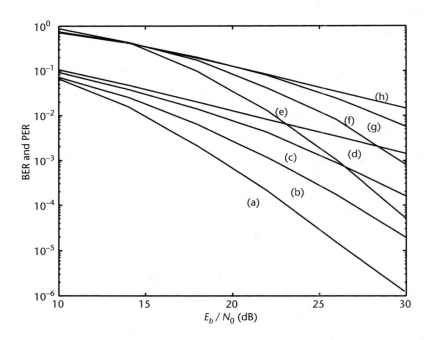

Figure 6.28   (a–d) BER and (e–h) PER versus mean E_b/N_0 for delay spreads of (a) and (e) 50 ns, (b) and (f) 20 ns, (c) and (g) 10 ns, and (d) and (h) 0.

6.5.5 Large Code Lengths

For OFDM systems with a large number of SCs, it may not be feasible to generate a sufficient number of complementary codes with a length equal to the number of channels. To avoid this problem, the total number of channels can be split into groups of channels; applying a complementary code to each group of subchannels increases the coding rate, at the cost of lessened error-correction capability and reduced PAP ratio. For 32 channels, for instance, 18 bits per symbol could be encoded using 8-PSK complementary codes. These codes would have a PAP ratio of 3 dB and a distance of 16 channel symbols, so 7 erroneous channels or 15 erased channels could be corrected. Instead of 32-channel codes, it is also possible to use four 8-channel codes or some other combination of shorter length codes. The sum of four 8-channel codes gives a total of 48 bits per symbol and a PAP ratio of 9 dB (6-dB reduction), while it is possible to correct one error or three erasures per group of eight channels.

6.6 Symbol Scrambling

Symbol scrambling techniques to reduce the PAP ratio of a transmitted OFDM signal can be seen as a special type of PAP reduction code. Symbol scrambling does not, however, try to combine FEC and PAP reduction such as is done by the complementary codes. The basic idea of symbol scrambling is that for each OFDM symbol, the input sequence is scrambled by a certain number of scrambling sequences. The output signal with the smallest PAP ratio is transmitted. For uncorrelated scrambling sequences, the resulting OFDM signals and corresponding PAP ratios will be uncorrelated, so if the PAP ratio for one OFDM symbol has a probability p of exceeding a certain level without scrambling, the probability is decreased to p^k by using k scrambling codes. Hence, symbol scrambling does not guarantee a PAP ratio below some low level; rather, it decreases the probability that high PAP ratios will occur. Scrambling techniques were first proposed in [19, 20] under the names selected mapping and partial transmit sequences. The difference between the two is that the first applies independent scrambling rotations to all SCs, while the latter only applies scrambling rotations to groups of SCs.

Figure 6.29 shows OFDM spectra for 64 SCs where the backoff is adjusted to maintain a −30-dB bandwidth that is twice the −3-dB bandwidth. A perfect linear power amplifier model is used, which clips the signal when the output power exceeds the saturation power level. The effect of scrambling has been simulated by scrambling the IFFT input data for each OFDM symbol with a certain number of independent complementary sequences and selecting the output symbol that gives the smallest PAP ratio. We can see from Figure 6.29 that scrambling with 1 and 10 codes gives rather small improvements of 0.25 and 0.75 dB in the required backoff, respectively, compared with the case without scrambling.

Figure 6.30 shows spectra similar to those presented in Figure 6.29, but for the stricter requirement of a −50-dB bandwidth that is twice the −3-dB bandwidth. In this case, scrambling gives more gain in the required backoff of up to 2 dB for 10

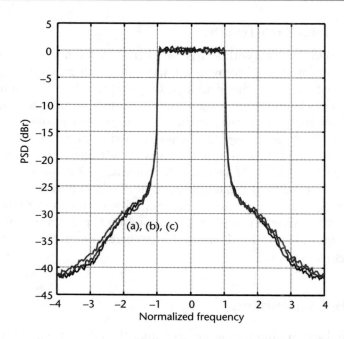

Figure 6.29 OFDM spectra for 64 SCs and $p=100$ using (a) no scrambling with 5.0-dB backoff, (b) 1 scrambling code with 4.7-dB backoff, and (c) 10 scrambling codes with 4.25-dB backoff.

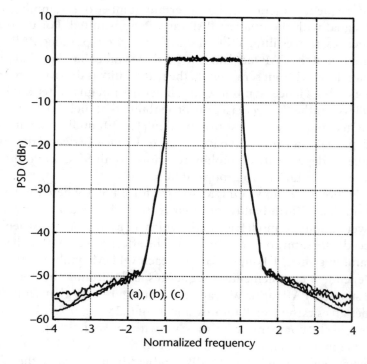

Figure 6.30 OFDM spectra for (a) no scrambling with 8.5-dB backoff, (b) 1 scrambling code with 7.2-dB backoff, and (c) 10 scrambling codes with 6.5-dB backoff.

scrambling codes. This is caused by the fact that with scrambling, the probability of exceeding a PAP ratio of 7 dB is much less than the probability of exceeding 4 dB (whose probability is close to one). As a result of this, for a backoff value of 4 or 5 dB, the amount of clipping interference is not much different from that without scrambling.

Figures 6.29 and 6.30 assume a perfectly linear power amplifier. In reality, however, the amplifier has a certain nonlinear TF. Figure 6.31 shows simulated spectra using Rapp's power amplifier model with nonlinearity parameter $p=2$, which closely resembles practical RF power amplifiers. We can see that the amplifier model changes the shape of the spectrum, but the relative gain of scrambling does not change significantly. Because for wireless systems the –30-dB bandwidth requirement of Figures 6.29 and 6.31 is more realistic than the –50-dB requirement of Figure 6.30, we can conclude that the benefits of scrambling are rather limited.

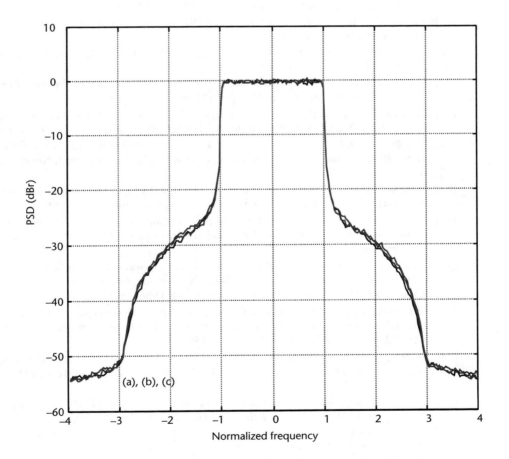

Figure 6.31 OFDM spectra for 64 SCs and Rapp's amplifier model with $p=2$ using (a) no scrambling with 5.8-dB backoff, (b) 1 scrambling code with 5.3-dB backoff, and (c) 10 scrambling codes with 5.2-dB backoff.

References

[1]   Pauli, M., and H. P. Kuchenbecker, "Minimization of the Intermodulation Distortion of a Nonlinearly Amplified OFDM Signal," *Wireless Personal Communications,* Vol. 4, No. 1, January 1997, pp. 93–101.

[2]   Rapp, C., "Effects of HPA-Nonlinearity on a 4-DPSK/OFDM Signal for a Digital Sound Broadcasting System," *Proc. of 2nd European Conference on Satellite Communications,* Liège, Belgium, October 22–24, 1991, pp. 179–184.

[3]   De Wild, A., "The Peak-to-Average Power Ratio of OFDM," M.Sc. thesis, Delft University of Technology, Delft, the Netherlands, September 1997.

[4]   May, T., and H. Rohling, "Reducing the Peak-to-Average Power Ratio in OFDM Radio Transmission Systems," *Proceedings of IEEE VTC'98,* Ottawa, Canada, May 18–21, 1998, pp. 2774–2778.

[5]   Li, X., and L. J. Cimini, "Effects of Clipping and Filtering on the Performance of OFDM," *Proc. of IEEE VTC'97,* 1997, pp. 1634–1638.

[6]   Wilkinson, T. A., and A. E. Jones, "Minimization of the Peak-to-Mean Envelope Power Ratio of Multicarrier Transmission Schemes by Block Coding," *Proc. of IEEE Vehicular Technology Conference,* Chicago, IL, July 1995, pp. 825–829.

[7]   Golay, M. J. E., "Complementary Series," *IRE Transactions on Information Theory,* Vol. IT-7, April 1961, pp. 82–87.

[8]   Sivaswamy, R., "Multiphase Complementary Codes," *IEEE Trans. on Information Theory,* Vol. IT-24, No. 5, September 1978, pp. 546–552.

[9]   Frank, R. L., "Polyphase Complementary Codes," *IEEE Trans. on Information Theory,* Vol. IT-26, No. 6, November 1980, pp. 641–647.

[10] Popovic, B. M., "Synthesis of Power Efficient Multitone Signals with Flat Amplitude Spectrum," *IEEE Trans. on Communications,* Vol. 39, No. 7, July 1991, pp. 1031–1033.

[11] Van Nee, R. D. J., "OFDM Codes for Peak-to-Average Power Reduction and Error Correction," *IEEE Global Telecommunications Conference,* London, England, November 18–22, 1996, pp. 740–744.

[12] Van Nee, R. D. J., "An OFDM Modem for Wireless ATM," *IEEE Symposium on Communications and Vehicular Technology,* Ghent, Belgium, October 7–8, 1996.

[13] Davis, J. A., and J. Jedwab, "Peak-to-Mean Power Control and Error Correction for OFDM Transmission Using Golay Sequences and Reed-Muller Codes," *Electronics Letters,* Vol. 33, 1997, pp. 267–268.

[14] Urbanke, R., and A. S. Krishnakumar, "Compact Description of Golay Sequences and Their Extensions," *Proc. of 34th Annual Allerton Conference on Communication, Control and Computing Pagination,* Urbana, IL, October 2–4, 1996, pp. 693–702.

[15] Urbanke, R., and A. S. Krishnakumar, "Compact Description of Golay Sequences and Their Extensions," *Lucent Technologies Technical Memorandum,* Doc. No. BL011217-961204-28TM, December 20, 1996.

[16] Davis, J. A., and J. Jedwab, *Peak-to-Mean Power Control in OFDM, Golay Complementary Sequences and Reed-Muller Codes,* HP Laboratories Technical Report, HPL-97-158, December 1997.

[17] IEEE, Draft Supplement to Standard Part 11: Wireless LAN MAC and PHY Specifications: Higher Speed PHY Extension in the 2.4 GHz Band, P802.11B/D6.0, May 1999.

[18] Grant, A., and R. van Nee, "Efficient Maximum Likelihood Decoding of Q-ary Modulated Reed-Muller Codes," *IEEE Communications Letters,* Vol. 2, No. 5, May 1998, pp. 134–136.

[19] Müller, S. H., et al., "OFDM with Reduced Peak-to-Average Power Ratio by Multiple Signal Representation," *Annals of Telecommunications,* Vol. 52, Nos. 1–2, February 1997, pp. 58–67.

[20] Müller, S. H., and J. B. Huber, "OFDM with Reduced Peak-to-Average Power Ratio by Optimum Combination of Partial Transmit Sequences," *Electronics Letters,* Vol. 33, No. 5, February 1997, pp. 368–369.

CHAPTER 7

A Novel Hybrid OFDM Concept

7.1 Introduction

This chapter proposes a novel hybrid OFDM/CDMA/SFH scheme [1–7]. Section 7.2 first presents a survey of existing modulation schemes providing multiple-access capability. After discussing the advantages and disadvantages of these schemes along with frequency-time diagrams, it introduces the proposed hybrid OFDM/CDMA/SFH scheme.

Thereafter, for the sake of continuity and to familiarize the reader with direct-sequence code division multiple access (DS-CDMA) systems, the IS-95 system is discussed. The various FH schemes and their advantages and disadvantages are examined. Finally, the proposed hybrid schemes is reviewed and how it corrects the disadvantages of the earlier discussed schemes is demonstrated.

Section 7.3 compares the hybrid system to the well-known MC-CDMA system. It makes a fine distinction proposed by Yee and Linnartz [8] and then traces its evolution to the OFDM-CDMA scheme as proposed by Fazel and Papke [9]. It then discusses the MC-CDMA system and derives a mathematical expression for its operation. A similar mathematical expression is derived for the proposed hybrid system, and they are shown to be mathematically similar. However, in real terms there is a lot of difference between the two techniques. This section concludes with a detailed comparison of both the techniques, giving their advantages and disadvantages. Finally, a qualitative comparison is made in terms of the computational power required for the two systems.

Section 7.4 studies DS-CDMA, DS-slow frequency hopping (SFH), and OFDM-CDMA-SFH (hybrid) systems analytically. Initially, it compares DS-CDMA and DS-SFH. In order to make a fair comparison, it uses noncoherent modulation [differential phase shift keying (DPSK)] in both cases, followed by a comparison using coherent modulation (QPSK). Thereafter, it compares DS-CDMA with another hopping scheme, namely OFDM-CDMA-SFH, based on another coherent modulation, namely 16-QAM. Finally, it carries out Monte Carlo simulations in the AWGN channel between OFDM, OFDM + CDMA, and OFDM + CDMA + SFH (hybrid) systems.

Section 7.5 first examines the manner of formulating the channel model. Then it discusses the approach toward the analytical evaluation of the BER, followed by a discussion of the simulation model. Finally, the section analyzes simulation results assuming perfect channel estimation and perfect synchronization.

Section 7.6 presents simulation results based on realistic channel estimation. Toward this end, it adopts the transform-domain processing (TDP) technique as a means to channel estimation. It then defines the OFDM system parameters for both the QPSK and 16-QAM modulations. Thereafter, simulation results are discussed based on the OFDM model. Finally, Section 7.7 presents the conclusions drawn.

7.2   Detailed Structure of Various Multiple-Access Schemes

This section discusses the various multiple-access schemes along with frequency-time diagrams and examines in detail the proposed scheme with its advantages over existing schemes.

7.2.1   Overview of Various Modulation Schemes

One of the candidates for a large data-rate modulation scheme is OFDM. In such systems very high data rates are converted to very low parallel-data rates using a series-to-parallel (S/P) converter. This ensures flat fading for all of the SCs; that is, a wideband signal becomes a packet of narrowband signals. This will automatically combat multipath effects, removing the need for equalizers and RAKE receivers. A variant of this approach was earlier introduced as MC-CDMA. This proposal envisages interfacing a DS-CDMA system with a system of orthogonal coding using Walsh coding [8–11]. We briefly examine the advantages and disadvantages of the prevailing systems before proceeding to examine the new proposal. It is pointed out that the advantages and disadvantages listed are not comprehensive, but only those relevant to this topic.

Figure 7.1(a) shows the frequency-time diagrams of the four types of access schemes. In Figure 7.1(a), the CDMA and CDMA-SFH diagrams are self-explanatory. The MC-CDMA and OFDM-CDMA, however, require some explanation.

- *MC-CDMA*: Suppose there is an 8-bit data "word." This word is then fed to a S/P converter. After this, each parallel bit is subjected to an 8-bit Walsh spreading using the code for that particular user in the FD. After this step, the signal is BPSK modulated on SCs F/T_b apart, where F is an integer number. If $F = 1$, we have a system similar to OFDM (but we achieve this without an OFDM modulator). These eight rows of the spread signal are then summed and transmitted.

- *OFDM-CDMA*: This technique is an improvement over the MC-CDMA approach. In this case, it is similar to MC-CDMA until the eight-bit Walsh spreading. After this, the spread signal is not summed, but subjected to an 8-point OFDM modulation. This means that we need to transmit 8 OFDM symbols before we are in a position to say that the eight-bit CDMA "word" has been completely transmitted. Alternately, if we use eight parallel IFFT boards, we can P/S convert the OFDM symbols and then transmit at the same time. In such a case, the frequency-time diagrams of MC-CDMA and OFDM-CDMA will be identical. However, this will be hardware intensive. In the receiver, an FFT is carried out, followed by RAKE combiners for summing.

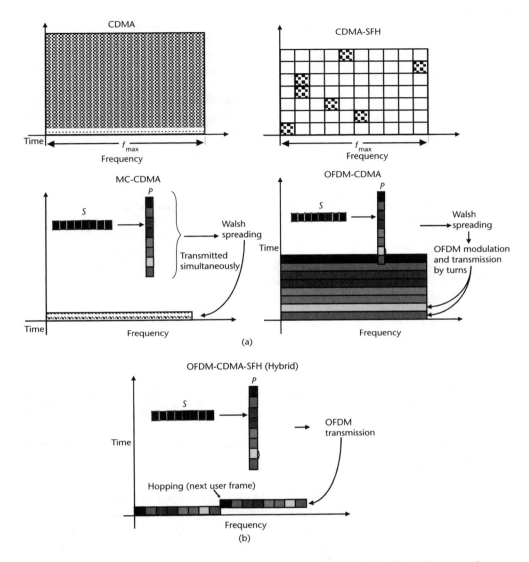

Figure 7.1   (a) Frequency-time diagrams, and (b) the OFDM-CDMA-SFH (hybrid) approach.

Table 7.1 shows that each multiple-access approach has its advantages and disadvantages. Note especially that one of the methods of implementing an MC-CDMA system is to adopt the OFDM/CDMA approach. This is still a complex procedure because it involves spreading each bit in a parallel manner using Walsh coding. There is an urgent need to develop a system that does the same thing, but in an easily achievable manner. This chapter pertains to such a system. This topic proposes a comprehensive approach maximizing the merits and minimizing the demerits of the individual schemes. This technique was first proposed during the 1999 ACTS summit [1]. In view of the novelty of the idea, as compared to MC-CDMA, this approach has been called, the hybrid OFDM/CDMA/SFH approach, or the

Table 7.1 Types of Multicarrier Access Schemes

Type	Advantages	Disadvantages	References
DS-CDMA	1. Ability to address multiple users simultaneously and at same frequency 2. Interference rejection	1. Problems due to near-far effect. 2. Complex TD RAKE receivers. 3. More difficult synchronization within a fraction of chip time	[1, 10, 12]
SFD-CDMA	1. Reduced near-far effect 2. Easier synchronization within a fraction of hop time 3. No need for contiguous bandwidths	1. Difficult coherent demodulation due to phase relationship during hops	[1, 10, 12]
MS-CDMA	1. Higher number of users as the full bandwidth is utilized unlike in DS-CDMA 2. Effective combination of all of the signal energy in the FD, unlike CDMA	1. Peak-to-average ratio problems 2. Synchronization problems 3. Overcrowding of the spectrum as each bit is spread across the available bandwidth based on Walsh coding 4. Complex FD RAKE receivers	[8, 9, 10]
OFDM-MA	1. Robust against multipath effects 2. Robust against narrowband interference 3. Capable of single-frequency operation	1. Sensitivity to frequency offset and phase noise 2. Synchronization problems 3. Large PAP ratios	[13]

hybrid approach for short. It has essentially been developed for the 60-GHz frequency, but is equally applicable at any other frequency provided the necessary bandwidth is available.

Before proceeding further, we shall briefly discuss the frequency-timing diagram of this new proposal.

OFDM-CDMA-SFH (Hybrid)   This is a new approach. In this case, suppose we have an eight-bit CDMA signal ("word"). We feed this word to a S/P converter. After this step comes a crucial difference. We do not carry out any Walsh spreading, but we straightaway carry out an 8-point IFFT (OFDM modulation). This means that one OFDM symbol equals the entire 8-bit CDMA word that we need to transmit. Hence, our information rate is eight times faster than in the previous two procedures. At the receiver, we carry out an FFT, but with another crucial difference: There is no need in our case to use RAKE combiners because we are not carrying out Walsh spreading in the FD in the interests of frequency diversity. On the contrary, we will accept the risk that some SCs will be in deep fade and correct for this eventuality using FEC coding (COFDM) or interleaving. This is different from the OFDM-CDMA approach discussed earlier, wherein RAKE combiners are used after OFDM demodulation in order to take advantage of the entire frequency spread of that particular bit. In reality, if we use an N-point OFDM system, we will need to use an N-finger RAKE combiner. This is extremely costly and, hence, a compromise is achieved by using fewer of fingers, for example, a 7-finger combiner. This means that we do not take advantage of the entire frequency spread anyway. In a way, this is a waste of resources. Therefore, after FFT we carry out a P/S conversion and then despread the CDMA signal. For user separation, we can use Walsh spreading in the transmitter, but this is done before the S/P converter in the transmitter, that is, in the TD. In the receiver we despread the Walsh signal after the P/S converter. This will ensure user separation. Other spin-offs from this technique are discussed in this chapter. Finally, in the frequency-time diagram, we have shown slow FH of the entire symbol. The hopping techniques are discussed further in this chapter.

This proposal pertains to the downlink as well as to the uplink, the only difference being that for the synchronous downlink we can apply orthogonal Walsh-Hadamard sequences leading to the well-known user separation for MC-CDMA systems. On the other hand, in the asynchronous uplink scenario, pseudonoise sequences are used with the drawback of high multiple-access interference (MAI).

7.2.2   DS-CDMA

Figure 7.2 shows a typical DS-CDMA system [12]. This is the IS-95 system. We shall briefly discuss the DS-CDMA schematic. The schematic pertains to the forward CDMA channel. The forward CDMA channel consists of 1 pilot channel, 1 synchronization channel, up to 7 paging channels, and up to 63 forward-traffic channels. The pilot channel allows a mobile station to acquire timing for the forward CDMA channel, provides a phase reference for coherent demodulation, and provides each mobile with a means for signal strength comparisons between base stations for determining when to handoff. The synchronization channel broadcasts synchronization messages to the mobile stations and operates at 1,200 bps. The paging channel is used to send control information and paging messages from the base station to the mobiles and operates at 9,600, 4,800, and 2,400 bps. The forward traffic channel supports variable user data rates at 9,600, 4,800, 2,400, or 1,200 bps. Data on the forward-traffic channel is grouped into 20-ms frames. The

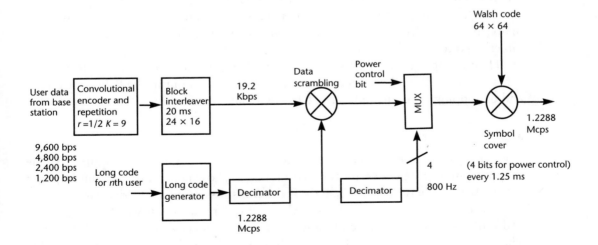

Figure 7.2   DS-CDMA schematic.

user data is first convolution coded and then formatted and interleaved to adjust for the actual user data rate, which may vary. Then the signal is spread with a Walsh code (which constitutes the inner code) and a long PN sequence at a rate of 1.2288 Mcps. The user data from the base station is input from a speech encoder. The speech encoder exploits pauses and gaps in speech and reduces its output from 9,600 bps to 1,200 bps during silent periods. In order to keep a constant baseband symbol rate of 19.2 Kbps, whenever the user rate is less than 9,600 bps, each symbol from the convolution encoder is repeated before block interleaving. The repetition rate is proportional to the input information rate. The convolution encoder is a half-rate encoder with a constraint length of nine.

After convolution coding and repetition, symbols are sent to a 20-ms block interleaver, which is a 24×16 array. In the forward channel, direct sequence is used for data scrambling. The long PN sequence is uniquely assigned to each user and is a function of the mobile station's electronic serial number (ESN) and its mobile station identification number (MIN). The PN sequence is generated at a rate of 1.2288 Mbps and is decimated to 19.2 Kbps. The data scrambling is then performed by modulo-2 addition of the interleaver output with the decimator output symbol. This is then multiplexed with four bits for power control and given an orthogonal covering using Walsh code. The Walsh code constitutes the inner coding. This is necessary because the PN sequence is in practice not enough to ensure channel isolation. The Walsh function matrix is a 64×64 matrix. Each row in this matrix corresponds to a channel number. For a channel number n, the symbols in the transmitter are spread by the 64 Walsh chips in the nth row of the Walsh function matrix. Channel 0 is the pilot channel, and channel 32 is the synchronization channel. The paging channels are assigned the lowest code channel numbers. The remaining channels are for forward traffic.

The reverse coding channel is similar, except that the Walsh function is used for data modulation instead of spreading to denote a particular channel. Also, in the reverse channel, there is no repetition unit as the mobile does not transmit during the repetition interval to avoid disturbing the other mobiles in the area.

The final output of the forward channel is bits emerging at 1.2288 Mbps. Normally, in DS-CDMA systems this is then transmitted. In our case, these bits are directly fed to the OFDM unit, which treats this as a binary data stream.

Having interfaced CDMA unit with OFDM, we shall now examine how to achieve slow FH for this system. The hopping is carried out in the RF upconverter, and dehopping is conducted in the RF downconverter.

7.2.3 SFH Interface

Slow FH is initiated in the RF upconverter. The principle is shown in Figure 7.3. The output from the D/A converter is fed to a mixer as shown in Figure 7.3. Our hopping rate is Δt ms, where Δt is the length of the transmission block as discussed below. This means that every Δt ms, we switch to a new frequency. If we assume a bandwidth of 100 MHz, we switch to another center frequency that is 100 MHz away. The hop set can be any number and is limited only by the extent of the available bandwidth. The RF downconverter is shown in Figure 7.4.

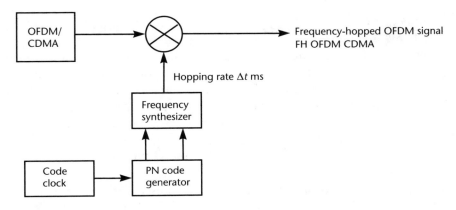

Figure 7.3   RF upconverter.

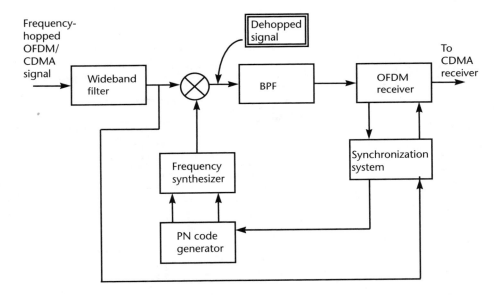

Figure 7.4   RF downconverter.

If the frequency pattern in the receiver synthesizer is synchronized with the frequency pattern of the received signal, then the mixer output is a dehopped signal at a fixed-difference frequency. Before demodulation, the dehopped signal is applied to a conventional RF receiver, that is, an RF amplifier, mixer, and so forth. In FH, whenever an undesired signal occupies a particular hopping channel, the noise and interference in that channel are translated in frequency so that they enter the demodulator. Thus, it is possible to have collisions in an FH system where an undesired user transmits in the same channel at the same time as the desired user. Normally, the type of modulation used in FH is noncoherent FSK as is usually done in DS-CDMA-SFH systems and the possible instantaneous frequencies change with each hop. In DS-CDMA-SFH systems, the 1.2288-MHz bit rate is frequency hopped at bit level (i.e., 1 hop/bit). Hence, coherent modulation techniques become difficult, and noncoherent FSK is preferred. In some instances, authors have gone to extremely slow hopping rates in order to achieve coherent modulation and demodulation. In our case, however, the FH is done on a frame-by-frame basis; that is, within a frame we can use coherent techniques like M-quadrature amplitude modulation (QAM) and the like. Hence, this is a superior approach as becomes readily apparent in the section on simulations (see Section 7.4). In this sense, our FH philosophy is essentially an SFH-TDMA approach wherein we have an n – TDMA system, where n is the number of available bandwidths for FH. The schematic is shown in Figure 7.5(a).

The shaded portion in Figure 7.5(a) represents the entire available bandwidth for an OFDM channel, for example, 100 MHz (Δf = 100 MHz). Within this bandwidth there are SCs. The sum of the SC bandwidths equals 100 MHz. The maximum bandspread is an integer multiple of 100 MHz, say, 4×100 = 400 MHz. This means

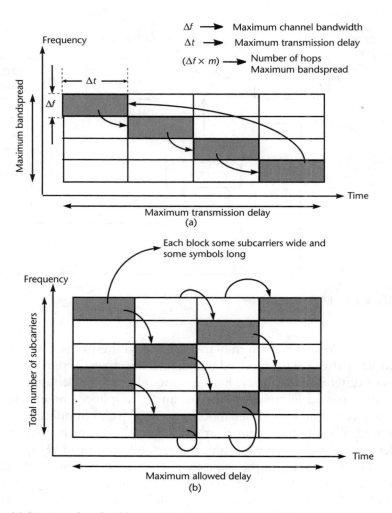

Figure 7.5   (a) FH across bandwidths, and (b) FH within a bandwidth.

that we can carry out four hops per user. The maximum transmission delay is the time allowed for the radio link (e.g., 20 ms). The Δt is the length depending upon the number of symbols being transmitted in one hop. As the next chapter will show, we have determined that 20 hops might be sufficient, but it is better to have more. Figure 7.5(a) demonstrates that in order to apply this scheme, we require very large bandwidths that are integer multiples of the basic signal bandwidth. This kind of bandwidth availability is presently difficult to ensure, unless we operate at 60 GHz.

An alternate, but more complex, approach is to keep the overall bandwidth constant, say at 100 MHz, and to hop within this bandwidth in a random manner. The choice of transmission block sizes in such a case becomes critical. A large block means more pilot signals for channel estimation, but it can also compromise the independence of the blocks. By independence, we mean the distance in time and the distance in frequency between two blocks, which should be much more than the coherence time and the coherence bandwidth, respectively, of the channel. If the block is too large, this distance can be compromised, and the number of independent blocks available for interleaving is lessened. This compromises channel diversity. For a fixed code, the system diversity depends upon both the number of independent blocks used for interleaving and the size of each block. This is a function of the data rate, the number of users we can handle at a time, and the maximum transmission delay. Hence, this is a trade-off. The reader is advised to refer to the study conducted by Telia Research on this subject as part of their UMTS proposal [14–16]. Figure 7.5(b) explains their approach.

Therefore, each CDMA user will be subjected to n frequency hops for frequency diversity and interference diversity (which lowers the required SNR and increases capacity). The frequency hop rate of an FHSS system is determined by the frequency agility of the receiver synthesizers, the type of information being transmitted, the amount of redundancy used to code against collisions, and the distance to the nearest potential interfering user. The near-far problem is, however, not totally avoided in FH systems because there will be some interference caused by stronger signals bleeding into weaker signals due to realistic filtering of adjacent channels. To combat occasional hits, error-correcting coding is required on all transmissions. By applying strong Reed-Solomon or other burst error-correcting codes, performance can be increased dramatically, even with occasional collisions.

7.2.4   OFDM/CDMA/SFH System Description

Figure 7.6 shows the overall concept discussed so far.

Brief Description   Figure 7.6 shows that the transmitter and receiver are each divided into three subsections:

Data modulation (demodulation) section;
DS-CDMA section;
SFH section.

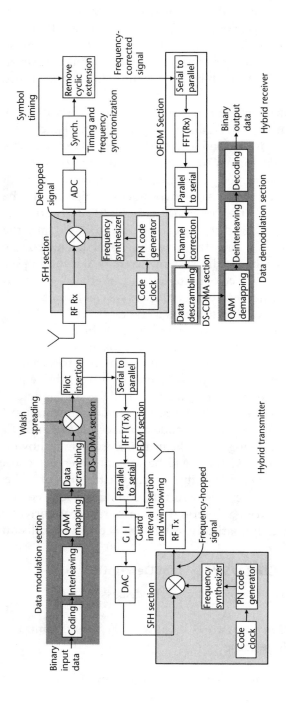

Figure 7.6 Overall system schematic.

The binary input data enters the data modulation section, where it is encoded by an FEC code. We can also use concatenated coding comprising a convolutional coding as an inner code, followed by an outer coding as a block code (e.g., a Reed-Solomon code) [13]. This makes for a large coding gain with less implementation complexity as compared to a single code. This coding is followed by interleaving to randomize the occurrence of bit errors due to deep fades across certain SCs. This is followed by QAM mapping. Thereafter, the data enters the DS-CDMA section, where the data is subjected to scrambling based on PN sequences. This aspect is similar to the implementation in Figure 7.2 for the IS-95 system. Thereafter, Walsh coding is used to provide orthogonal covering because PN sequences by themselves are insufficient to ensure user separation. The Walsh function matrix will be a $N \times N$ matrix where N is the number of OFDM points. We then obtain what we can call a DS-CDMA signal. We insert pilot symbols after this step. In doing so, we must take care regarding the size of the Walsh matrix. For example, if our OFDM system has 32 points, we can have at the most a 24-length Walsh sequence. This leaves eight SCs for pilots, assuming that we utilize all of the SCs. In practice, this is not possible because we need to leave the edge SCs unutilized. This problem is explained as follows. We need to allow for the skirt of the lowpass antialiasing filter in the receiver. SCs that lie beyond the bandwidth of interest should contain zero information because these SCs will lie along the slope of the lowpass filter. If they contain information then their amplitudes will not be uniform since they lie along the slope. Recall that the fundamental assumption for orthogonality between SCs is that they have constant amplitude and differ only in phase. Hence, if any SC of interest lies along the slope of the lowpass filter, we will have ICI. Therefore, we need to allow a safety zone, as it were, around each frame by inserting zeros into SCs around the edges of the OFDM symbol. During this process, we must ensure that all of the SCs of interest lie within the passband of the lowpass filter. Furthermore, the communication spectrum is crowded. This means that no extraneous signals should exist beyond the passband of the filter (i.e., along the slope). The zeros ensure this. The data sequence is then given to the OFDM section. This section is self-explanatory. One point to be noted here is that each user in the DS-CDMA section will share the entire lot of SCs with other users. Discriminating between users will only be possible in the DS-CDMA section of the receiver after descrambling and will be based on the orthogonality of the PN sequences and the Walsh coding. The analog signal coming from the DAC is then frequency-hopped in the SFH section, before being fed to the RF transmitter. The hop set for each user usually bears a definite relationship to the PN sequence of a particular user. During this process, ensure the following:

- That the frequency synthesizer of the hopper and the carrier beat frequency oscillator of the RF transmitter are phase locked;
- That the frequency synthesizer of the dehopper and the beat frequency oscillator of the receiver (whose operating frequency is controlled by the synchronization circuit) are also phase locked.

Failure to ensure these two aspects will result in ICI.

The hybrid receiver uses exactly the reverse operation. The dehopped signal is given to the ADC and, thereafter, the digital-signal processing starts with a training phase to determine the symbol timing and frequency offset. An FFT is used to demodulate all of the SCs. The output of the OFDM section is the DS-CDMA sequence, which is then descrambled. The output of the DS-CDMA section is the QAM sequence, which is mapped onto binary values and decoded to produce binary output data. In order to map the QAM values onto binary values successfully, first the reference phases and amplitudes of all SCs have to be acquired. Alternatively, differential techniques can be applied.

However, a few salient points should be noted:

1. *Bandwidth and other considerations:* In this chapter, we are assuming both Rayleigh and Rician fading conditions and AWGN. We are also assuming perfect OFDM synchronization with no carrier offset. Multimedia requirements of high bit rates, typically 155 Mbps, require wide bandwidths of around 100 MHz or higher. The CDMA system in Figure 7.2 pertains to a voice channel with a bandspread factor of 128 for a data rate that is at most 9,600 bps. This is necessary due to adverse transmission conditions and to there being a lot of users at that frequency (around 850 MHz). In our case, however, we intend to operate at around 60 GHz, where larger bandwidths are available. This entails Rician fading conditions and LOS transmissions, conditions that are not so severe. Therefore, a bandspread factor of 128 is unnecessary. More likely, a bandspread factor of 10 or less will prove sufficient. This, however, has to be verified by extensive simulations. If we assume a bandspread factor of 10, then we require a bandwidth of at most 1 GHz for a 100-Mbps data rate. In such a case the proposal shown in Figure 7.5(b) is a better approach. If the bandspread factor is much less than 10, we can adopt the approach in Figure 7.5(a), which is easier to implement.

   The basic motivation for the scheme in Figure 7.6 is not so much having a robust design, as having a design that can incorporate a lot of users. There is no need for such a robust design at such frequencies. However, Section 7.4 will show that this approach also yields low BERs suitable for the transmission of high data rates.

   We can expect a steep rise in the number of users when high data rates become realizable, especially with regard to videotelephones. The CDMA aspect (code diversity) really gives rise to a lot of users because FH has been introduced to obtain frequency diversity to reduce the near-far problem. This limits the number of users in order to avoid collisions. The CDMA aspect makes up for this limitation by introducing a larger number of users due to code diversity. Interleaving and error-correction coding may be dispensed with if the need so arises, that is, if the channel is not severe. In case the channel does pose problems in the foreseeable future, we can increase the spread factor of the CDMA transmission (increase the bandwidth) or introduce FEC and interleaving. The OFDM aspect is required because it eliminates the need for RAKE receivers (as compared to pure DS-CDMA systems because there are no multipath delay effects) and allows us to use coherent modulation even when frequency hopping (because we will now hop on an OFDM symbol basis), unlike in most SFH systems wherein maintaining phase continuity during hopping is difficult. It also helps reduce the burden of synchronization related to CDMA systems (see list item 5, synchronization). In this connection, MC-CDMA also uses OFDM techniques, but with RAKE receivers (in the FD, due to Walsh spreading). In our style of signal processing, we do not use RAKE receivers. This crucial change from the MC-CDMA design results in a massive saving of hardware. We will accept the risk that some SCs will be in deep fade and correct for this eventuality using FEC coding (COFDM) or interleaving. This is different from the OFDM-CDMA approach discussed earlier, wherein RAKE combiners are used after OFDM demodulation in order to take advantage of the entire frequency spread of that particular bit. In reality, if we use an N-point OFDM system, we will need to use an N-finger RAKE combiner. This is extremely costly; hence, a compromise is achieved by using fewer fingers (e.g., a 7-finger combiner). This means that we do not take advantage of the entire frequency spread anyway. In a way, this is a waste of resources. Taking such matters into consideration, the hybrid system does not spread each bit in the FD and does not, therefore, use RAKE combiners. Therefore, we call this approach the hybrid OFDM/CDMA/SFH approach and not MC-CDMA.

2. *Coding:* In the CDMA transmitter (as in an IS-95 system), there are two levels of coding: convolution encoding (for error correction) or concatenated coding and Walsh encoding (this is a spreading code, not an error-correction code). The latter is an orthogonal coverage because PN sequences by themselves are insufficient to ensure channel isolation. The Walsh coding ensures orthogonality between users. The convolutional encoding ensures robustness of data.

3. *Modulation:* Unlike in a pure DS-CDMA system, in our case, the CDMA sequence after Walsh coding does not get converted to RF, but instead is fed as an input to the OFDM transmitter. In the OFDM transmitter it gets modulated as an OFDM signal, then via a P/S converter gets converted to RF.

4. *CDMA receiver:* Similarly, the OFDM receiver gives the CDMA receiver a sequence at chip rate after OFDM demodulation. Thereafter, CDMA signal processing is carried out in that there is a digital correlator that ensures channel isolation based on PN sequences and Walsh coding. The output of the correlator is then given to a Viterbi decoder (for convolution decoding). The output from this decoder is the required data sequence. RAKE receivers are not required in this case, unlike in DS-CDMA systems, because the OFDM system has no deleterious effects due to multipath.

5. *Synchronization:* Stringency of synchronization is, however, still required as the PN sequences need to be synchronized. However, in such a hybrid system, the burden of synchronization is transferred to the OFDM system. The OFDM synchronization system is more sophisticated than CDMA

synchronization systems as the OFDM system utilizes the cyclic prefixes for synchronization. Hence, the PN sequences emerging from the OFDM system and going to the CDMA system are already better synchronized than in a pure CDMA system. The reader will recall that synchronization is one of the limiting factors in CDMA systems for high data rates. It is expected that in our system, such problems will be considerably reduced, especially in the uplink because the mobile receivers, thanks to OFDM, will be better synchronized to the transmitter. This will reduce MAI as compared to a pure DS-CDMA system.

6. *Bit-error probabilities:* The proposed system is essentially a CDMA/ OFDM-FH system because the transmission and reception are carried out by the OFDM-FH system. The CDMA aspect generates the data stream, but in a more complicated way. In Section 7.5, we compare the simulation results in AWGN between OFDM, OFDM/CDMA (MC-CDMA) and the hybrid system (OFDM/CDMA/SFH).

7. *Trade-off between OFDM and CDMA:* The OFDM-FH system by itself does not solve the multimedia requirement because multimedia requires very high bit rates, typically 155 Mbps. This requires large bandwidths of typically 100 MHz. By using FH among users to reduce the near-far effect suffered by CDMA systems, our number of users comes down drastically, as it is limited by the bandwidth available. By adding CDMA, we have rectified this problem by enhancing the number of users because CDMA supports additional users (being limited only by MAI) working at the same frequency. Hence, there is a trade-off, which we clarify using an example. Suppose in a hybrid system the CDMA end cannot handle more than 20 users due to MAI. These 20 users share one hop set. Therefore, among these users there will be adverse performance due to near-far effect. If we find that this near-far effect is intolerable, we reduce the number of users to, say, 10 and make the remaining 10 share another hop set. Due to bandwidth constraints, suppose we can utilize only two hop sets. Then, we once again have a total of only 20 users for this hybrid system. But on the other hand, if the near-far effect is not too serious for 20 users, we can assign the other hop set to another 20 users for 40 users in all. This trade-off between control of near-far effect and number of users depends upon channel conditions. The hybrid system gives us this flexibility. Hence, there is eventually a trade-off between our desire to control the near-far effect and the number of users we desire.

8. *CDMA signal processing:* It will be argued that the present data rate of 1.2288 Mbps in IS-95 systems is woefully inadequate for multimedia applications. This is acknowledged. However, bear in mind that if we want a large number of users, we need to use CDMA techniques. Hence, efforts must be made to increase data rates using better phase locked loops (PLLs) for synchronization and high-speed digital electronics. In this proposed hybrid concept, the entire CDMA signal processing is carried out in the digital domain both for the transmitter as well as for the receiver.

7.2.5   Summary

This section has dealt with the problem of multimedia communications, stressing the need for high bit rates in the frequency spectrum of 60 GHz, especially as regards videotelephones. We made the following observations:

1. The MC-CDMA system, whether it is implemented as an OFDM/CDMA system or in any other manner, is extremely complex as it involves spreading each bit across the available bandwidth. Our approach, on the other hand, spreads the entire word before the S/P converter, thereby avoiding the complication of single-bit spreading as is done in MC-CDMA systems. This means that we spread in the TD, whereas MC-CDMA systems spread in the FD to achieve the same final result. The MC-CDMA system, moreover, cannot cater to SFH to reduce near-far effect because in such a case the system will become even more complex. In our approach, on the other hand, this can easily be done. Hence, we choose to call our approach the hybrid OFDM/CDMA/SFH approach, or the hybrid approach for short. This approach has not been suggested before, and it was first presented at the 1999 ACTS summit.

2. The advantage of the hybrid approach lies in its easy implementation as compared to MC-CDMA because the hardware is less complex. This aspect will be further examined in Section 7.3.

3. The problems of synchronization are transferred from the CDMA to the OFDM modulator. We expect that it will be easier to solve the synchronization of the OFDM modulator as compared to a CDMA modulator because the former uses cyclic prefixes that can be exploited to achieve synchronization.

   This proposal pertains to the downlink as well as to the uplink, the only difference being that for the synchronous downlink we can apply orthogonal Walsh-Hadamard sequences, leading to the well-known user separation for MC-CDMA systems. On the other hand, in the asynchronous uplink scenario, pseudonoise sequences are used with the drawback of high MAI.

Table 7.2 summarizes the overall system aspects.

7.3   Comparison to MC-CDMA

In this section, a comparison is made between the hybrid approach and the MC-CDMA system as proposed by Yee and Linnartz [8] and Fazel and Papke [9]. It shows that the proposed system is simpler to implement and, without SFH, performs similarly to MC-CDMA systems. It shows that the hybrid system has greater flexibility in terms of the size of the OFDM modulators and number of users and lends itself easily to FH with much higher data rates as compared to MC-CDMA systems.

Table 7.2 Hybrid System Overall Aspects

	Problem	Solution
1.	OFDM system only supports one user.	We use OFDM/CDMA, which supports multiple users.
2.	CDMA does not permit very high data rates owing to frequency-selective fading at high data rates.	OFDM counters this with S/P conversion, allowing flat fading at SC level.
3.	CDMA system suffers from the near-far effect in the uplink.	This is solved in the hybrid system using SFH.
4.	CDMA systems use DS-SFH to control the near-far effect, but DS-SFH mostly supports noncoherent modulation owing to hopping at bit level. Coherent modulation is possible, but maintaining phase coherence between hops is difficult.	OFDM-FH systems hop on a frame basis, allowing coherent modulation.
5.	CDMA systems cannot indefinitely support multiple users due to MAI and signal-to-interference (SI) problems due to too many users.	The hybrid system allows any number of users by increasing the number of hops. Hence, number of users = number per CDMA system number of hops. Bandwidth should, however, be available.
6.	CDMA poses synchronization problems at very high chip rates.	OFDM systems have fewer synchronization problems due to using cyclic prefixes.

7.3.1 Background

MC-CDMA was first proposed by Linnartz and Fettweis [17] in PIMRC '93. Their proposal as envisaged in [17] involved transmitting each data symbol over N narrowband SCs, where each SC is encoded with a 0 or π phase offset. If the number of and spacing between SCs is appropriately chosen, it is unlikely that all of the SCs will be in deep fade; consequently, frequency diversity is achieved. As an MC-CDMA signal is composed of N narrowband SC signals, each with a symbol duration T_b much larger than the delay spread T_d, an MC-CDMA signal will not experience significant ISI. Multiple access is achieved with different users transmitting at the same set of SCs, but with spreading codes that are orthogonal to the codes of other users. A paper by Fazel and Papke [9] on OFDM/CDMA improves considerably on the MC-CDMA proposal, although in mathematical terms it is similar to MC-CDMA, but more elegant to implement. From this point forward, when we say MC-CDMA, we mean OFDM-CDMA. This is different from OFDM-CDMA (hybrid), which is the new proposal, discussed in the previous section.

7.3.2 Basic Principles of MC-CDMA

This portion is based on the work done by Hara and Prasad [18–21]. An MC-CDMA transmitter spreads the original signal using a given spreading code in the FD. In other words, a fraction of the symbol corresponding to a chip of the spreading code is transmitted through a different SC. For MC transmission, it is essential to have frequency-nonselective fading over each SC. Therefore, if the original symbol rate is high enough to become subject to frequency-selective fading, the signal needs to be first S/P converted before spreading over the FD. The basic transmitter structure of the MC-CDMA scheme is similar to that of the OFDM scheme, the main difference being that the MC-CDMA scheme transmits the same symbol in parallel through a lot of SCs, while the OFDM scheme transmits different symbols.

Figure 7.7 shows the MC-CDMA transmitter for the jth user with coherent detection binary phase shift keying (CBPSK) format. The input information sequence is first converted into P parallel-data sequences. $(a_{j,0}(i), a_{j,1}(i),..., a_{j,P-1}(i))$ and then each S/P converter output is multiplied by the spreading code with length K_{MC}.

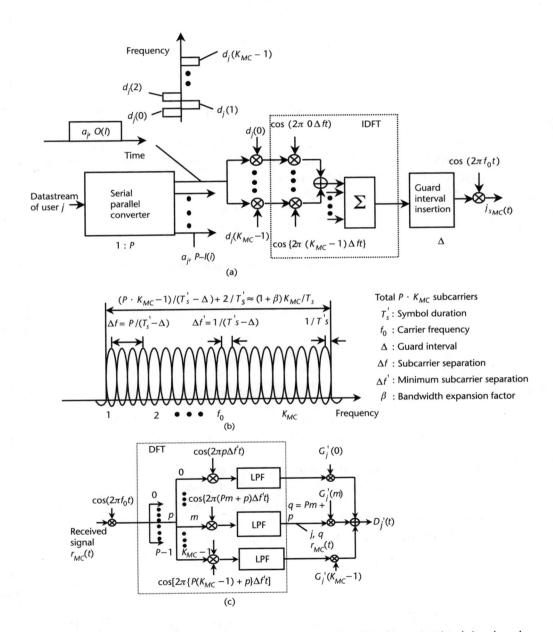

Figure 7.7 MC-CDMA system: (a) transmitter, (b) power spectrum of its transmitted signal, and (c) receiver.

All the data in total $N = P \times K_{MC}$ (corresponding to the total number of SCs) is modulated in baseband by the IFFT and converted back into serial data. The GI Δ is inserted between symbols to avoid ISI caused by multipath fading, and finally, the signal is transmitted after RF upconversion. The complex equivalent lowpass transmitted signal is written as

$$s_{MC}^{j}(t) = \sum_{i=-\infty}^{+\infty}\sum_{p=0}^{P=1}\sum_{m=0}^{K_{MC}-1} a_{j,p}(i)d_{j}(m)$$

$$\cdot p_{s}(t - iT'_{s})e^{j2\pi(Pm+p)\Delta f'(t-iT'_s)} \tag{7.1}$$

$$T'_{s} = PT_{s} \tag{7.2}$$

$$\Delta f' = 1 / (T'_{s} - \Delta) \tag{7.3}$$

where $\{d_{j}(0), d_{j}(1),...,d_{j}(K_{MC}-1)\}$ is the spreading code with length K_{MC}, T'_{s}, is the symbol duration at SC, $\Delta f'$ is the minimum SC separation, and $p_{s}(t)$ is the rectangular symbol pulse waveform defined as

$$p_{s}(t) = \begin{cases} 1, (-\Delta \leq t \leq T'_{s} - \Delta) \\ 0, otherwise \end{cases} \tag{7.4}$$

The bandwidth of the transmitted signal spectrum is written as [see Figure 7.7(b)]

$$\begin{aligned} B_{MC} &= (P \cdot K_{MC} - 1) / (T'_{s} - \Delta) + 2/T'_{s} \\ &\approx K_{MC} / T_{s} / (1 - \Delta / P) \\ &= (1 + \beta)K_{MC} / T_{s} \end{aligned} \tag{7.5}$$

$$\beta = \Delta / P, (0 \leq \beta \leq 1.0) \tag{7.6}$$

where β is the bandwidth expansion factor associated with the GI insertion.

Note that in (7.1) no spreading operation is done in the TD. Equation (7.2) shows that the symbol duration at SC level is P times as long as the original symbol duration due to S/P conversion. Although the minimum SC separation is given by (7.3), the SC separation for $a_{j,p}(i)$ is $\Delta f = P/(T_{s}'-\Delta)$. Therefore, when setting K_{MC} to one, the transmitted waveform given by (7.1) becomes the same as an OFDM waveform with P SCs.

On the other hand, the received signal is written as

$$r_{MC}(t) = \sum_{j=1}^{J}\int_{-\infty}^{+\infty} s_{MC}^{j}(t - \tau) \otimes h^{j}(\tau;t)d\tau + n(t)$$

$$= \sum_{i=-\infty}^{+\infty}\sum_{p=0}^{P-1}\sum_{m=0}^{K_{MC}-1}\sum_{j=1}^{J} z_{m,p}^{j}(t)a_{j,p}(i)d_{j}(m) \cdot p_{s}(t - iT'_{S})e^{j2\pi(Pm+p)\Delta f't} + n(t) \tag{7.7}$$

where $z_{m,p}^{j}(t)$ is the received complex envelope at the $(mP + p)$-th SC of the jth user.

The MC-CDMA receiver requires coherent detection to perform a successful despreading operation. This complicates matters in an already complicated schematic. Figure 7.7(c) shows the MC-CDMA receiver for the j'th user. After down-conversion, the mth SC component ($m = 0, 1, ..., K_{MC} - 1$) corresponding to the received data $a_{j',p}(i)$ is first coherently detected with FFT and then multiplied with the gain $G_{j'}(m)$ to combine the energy of the received signal scattered in the FD. The decision variable is the sum of the weighted baseband components given by (we can omit the subscription p without loss of generality)

$$D_{MC}^{j'}\left(t = iT_s'\right) = \sum_{m=0}^{K_{MC}-1} G_{j'}(m)y(m) \tag{7.8}$$

$$y(m) = \sum_{j=1}^{J} z_m^{j}(iT_s')a_j d_j(m) + n_m(iT_s') \tag{7.9}$$

where $y(m)$ and $n_m(iT_s')$ are the complex baseband component of the received signal after downconversion and the complex additive Gaussian noise at the mth SC at $t = iT_s'$, respectively. The system adopts different combining strategies to recover the signal. These are discussed in the references.

7.3.3   The Hybrid System

We now examine the equations pertaining to the hybrid system without SFH. From now on, we shall call this OFDM-CDMA (hybrid).

The transmitted signal of the ith user [22] is written as

$$s_{sp}^{i}(t) = \sum_{u=-\infty}^{+\infty} \sum_{n=0}^{P-1} \sum_{m=1}^{N_{PG}} \text{Re}\left[\begin{array}{c} b_n^{i}(u)c_m^{i}(u) \cdot p_c\left\{t - (m-1)T_c' - uT_s'\right\} \\ \cdot \exp\left\{j2\pi\left(f_0 + n\Delta f_{sp}\right)t\right\} \end{array}\right] \tag{7.10}$$

where

$b_n^{i}(u)(=1 \pm j, -1 \pm j)$ is the uth input data at the nth carrier of ith user after S/P conversion.

$c_m^{i}(u)$ is the mth chip of the spreading code for the ith user.

$T_s'(= PT_s)$ is the symbol duration, and $T_c'(=T_s'/N_{PG})$ is the chip duration after S/P conversion respectively.

f_0 is the lowest carrier frequency, and $\Delta f_{sp}(=1/T_c')$ is the carrier separation.

$p_c = 1(0 \le t \le T_c')$, $p_c = 0$ (otherwise).

In this system, the symbol period of T_s' becomes P times as long as T_s of the single-carrier case, and the influence of multipath fading is released.

From (7.10), the transmitted signal of the ith user at $u = 0$ is represented as

$$s(t) = \sum_{n=1}^{P} \sum_{m=1}^{N_{PG}} \mathrm{Re}\left[b_n c_m p_c(t')e^{j2\pi f_n t}\right] \tag{7.11}$$

where

$$p_c(t') = p_c\{t - (m-1)T'_c\}, f_n = f_0 + n\Delta f_{sp}$$

and i is not marked for simplicity. If the channel is considered unchanging for one symbol duration, the channel IR is expressed with the delta function $\delta(t)$ as

$$h(t) = \sum_{l=1}^{L} h_l \delta(t - \tau_l) \tag{7.12}$$

where h_l and τ_l are the amplitude and delay time of the lth path, respectively. The received signal is given by

$$\begin{aligned} r(t) &= (s \otimes h)(t) + \mathrm{Re}\left[n(t)e^{j2\pi f_{P/2}t}\right] \\ &= \sum_{l=1}^{L} \sum_{n=0}^{P-1} \sum_{m=1}^{N_{PG}} \mathrm{Re}\left[b_n c_m p_c(t)e^{j2\pi f_n t}\right] + \mathrm{Re}\left[n(t)e^{j2\pi f_{P/2}t}\right] \\ &= \sum_{n=0}^{P-1} \sum_{m=1}^{N_{PG}} \mathrm{Re}\left[H_k b_n c_m p_c(t)e^{j2\pi f_n t}\right] + \mathrm{Re}\left[n(t)e^{j2\pi f_{P/2}t}\right] \end{aligned} \tag{7.13}$$

where $(s \otimes h)(t)$ is the convolution of $s(t)$ and $h(t)$, and H_n is expressed by

$$H_n = \sum_{l=1}^{L} h_l e^{-j2\pi f_n \tau_l} \tag{7.14}$$

In the receiver, $r(t)$ is transferred by a local oscillator whose frequency is $f_{P/2}$. The resulting complex signal becomes [23–26]

$$r'(t) = \sum_{n=0}^{P-1} \sum_{m=1}^{N_{PG}} H_n b_n c_m p_c(t)e^{j2\pi \frac{n-P/2}{T'_s}} + n(t) \tag{7.15}$$

Therefore, the mth chip of the nth carrier after FFT is shown as

$$r_n = \frac{1}{T'_c} \int_0^{T'_c} r'(t)e^{-j2\pi \frac{n-P/2}{T'_s}} dt = H_n b_n c_m + n_n \tag{7.16}$$

7.3.4   Comments on the MC-CDMA Technique

We can see that the MC-CDMA technique is very complex. However, mathematically the BER derived from this approach is no different from the OFDM-CDMA (part of the hybrid proposal under consideration) approach. This is apparent if we compare the structure of (7.7) with (7.15). They differ only in the manner that the signal processing is implemented. The important distinction is that in MC-CDMA, if we have a set of OFDM symbols, say, eight, we implement an 8-point OFDM system. This means these eight symbols comprise eight OFDM frames. One OFDM frame comprises eight SCs. Each symbol uses one complete OFDM frame. In contrast, the hybrid approach transmits all of the eight symbols in one OFDM frame. Admittedly, the frequency-diversity effect is applied more rigorously in MC-CDMA as compared to the hybrid approach because in the former method each symbol is transmitted over all of the SCs; that is, the spreading is carried out over the FD as opposed to the TD in the hybrid method. This means that if there are certain carriers that do not perform well, we are still assured that the receiver will get that particular symbol through other carriers that do operate well. The implication here is that in fading channels in the presence of certain carriers in deep fade, we are likely to lose information on those SCs. This will not happen in MC-CDMA as the same bit is on all SCs. Hence, in such a case, MC-CDMA is likely to perform better. In fact Takeda et al. [23] extended this principle further by getting feedback from the receiver as to which carriers are in good condition and using only those carriers that have performed well. They call this approach, the partial bandwidth transmission (PBT) approach.

The approach to simulation is also the same in the hybrid as well as the MC-CDMA. In both cases, each symbol in the set of symbols to be transmitted is spread as per a code. In MC-CDMA the spreading code is based on Walsh-Hadamard coding; that is, each user is allotted one row of the Hadamard matrix. This ensures that the users do not clash because the rows are mutually orthogonal. In the hybrid approach, the spreading is carried out as per some PN sequence or Gold sequence. In both cases depending upon the spreading length, there will be processing gain. In the receiver, the received signal is once again multiplied by the same sequence. Thereafter, there is an important distinction between the two systems. In the hybrid system, the received symbols are summed (integrated) after multiplication, whereas in the MC-CDMA approach, this summing is carried out as per some combination law (e.g., maximal ratio combining or equal gain combining) because the MC-CDMA system utilizes RAKE combiners. Hence, the results obtained using the OFDM/CDMA (hybrid) approach (i.e., the hybrid approach without SFH) in this book apply equally well to MC-CDMA systems provided no SCs are in deep fade, leading to loss of information.

There is also one more point in favor of the hybrid approach. We have seen that if we choose, we can also use Walsh-Hadamard coding in the hybrid approach as an additional precaution in addition to PN sequences. This is carried out before S/P conversion.

Certain additional points in favor of the hybrid system include the following:

1. In the MC-CDMA approach, the size of the OFDM modulator is dependent upon the size of the Hadamard matrix. In the hybrid approach, we select the number of carriers and, consequently, the size of the OFDM system based on the channel condition. Thus, it has robustness against large propagation delay. The designer consequently has total flexibility in this respect.

2. In the MC-CDMA approach, the number of users depends upon the size of the Hadamard matrix. We cannot have additional users because there are no more rows left in the Hadamard matrix. In the hybrid system, on the other hand, the number of users can be increased indefinitely. This means that if the CDMA system cannot take more than, say, 20 users, we simply add another frequency hop to the system. The total number of users = number of users per CDMA channel × the total number of frequency hops. Bandwidth should, however, permit this.

3. FH is another major problem in MC-CDMA systems. Consider a case where we are transmitting a set of eight symbols based on a certain type of modulation like QPSK or 16-QAM. We use an 8-size Hadamard matrix and carry out an 8-point OFDM modulation. This means we need to transmit eight OFDM frames for the complete set. If we plan to frequency hop, then we need to delay the hopping until these eight frames are transmitted. If we do choose to hop with every frame as we do in hybrid systems, the task of compiling the eight symbols at the receiver becomes complex.

4. The data rate will become very slow in MC-CDMA systems as compared to hybrid systems. Once again, consider the case of an eight-symbol set based on any type of modulation like QPSK or 16-QAM. We need to transmit eight frames before the symbol set is considered to have left the transmitter. In contrast, the hybrid system transmits just one frame.

In order to estimate the processing power required to implement a practical multimedia system, consider an example.

Total bandwidth: 150 MHz;
User capacity: Single user (we shall hand over the multiuser problem to the CDMA part of the hybrid system);
Modulation used: 16-QAM;
FFT size: 512;
Guard period: 128 samples.

If we assume the number of active carriers is 200, we obtain the following parameters[1]:

Data rate: 600 Mbps;
Symbol duration: 1.3 μs;
Total frame time: 1.7 μs $\approx$ 2 μs.

We know that the number of complex calculations required for a 512-point FFT is 6,912. The maximum time that can be taken in performing the calculation is once every symbol, that is, once every 2 μs. If we assume that the processor used requires

---

1.   The remaining bits are zeros since we need to avoid using the carriers close to the Nyquist frequency because of the attenuating effects of the lowpass filters used in the DACs and ADCs. Furthermore, sensitivity to frequency offsets increases with size of the FFT. This in turn will require a high degree of frequency stability.

two instructions to perform a single complex calculation, and that there is an over-head of 30% for scheduling of tasks and other processing, the minimum processing power required for this is then

$$MIPS = \frac{6,923 \times 2}{2 \times 10^{-6}} \times 1.3 \times 10^{-6} = 8,986$$

Thus, the transmitter requires more than 9,000 MIPS in order to implement the transmitter. The receiver will require just as much. Thus, a full OFDM transceiver will require two boards capable of more than 9,000 MIPS each. This is beyond the range of current processors, but within the capability of hardware-based systems. The MC-CDMA system on the other hand will require 512 such IFFT boards in the modulator (a 200-bit word + 312 zeros = 512 bits) and as many FFT boards in the receiver. These boards will need to be in parallel. Alternatively, these boards will need to be replaced by a single card in modulator/demodulator, which is as fast. Such fast FFT boards are currently not available. This will make the system extremely expensive. Even in this case, the P/S converter will take the output of each FFT board in turn and serialize it. This means that 512 OFDM symbols or 512 OFDM frames, if we include the guard bands, will have to be transmitted before the system is ready to look at the next word (the MC-CDMA system is transmitting a 512×512 matrix). This cannot be avoided. This will slow down the overall data rate considerably as compared to a 512-point OFDM system, which is roughly 512 times faster, as it will transmit 512 bits at once as one OFDM symbol or frame (if we include the guard bands). This hardware count for the MC-CDMA system is further increased by the use of RAKE receivers.

7.3.5   Summary

In this section, we examined the mathematics and signal processing of both the MC-CDMA system as well as the proposed hybrid system as given in this book. We found basically that the MC-CDMA and the OFDM-CDMA (hybrid) (i.e., the hybrid system without SFH) systems are identical. The MC-CDMA system, however, proved more complex than the hybrid system without SFH for the same quality of performance. In particular the following were the salient observations:

1. The MC-CDMA system has no flexibility in the selection of the size of the OFDM modulator. We would prefer to choose the size of the OFDM modulator based on the channel condition, which consequently is robust against large propagation delay. The designer has total flexibility in this respect. In the MC-CDMA system, there is no such flexibility as the OFDM modulator size is dependent upon the size of the Hadamard matrix.

2. The number of users in the MC-CDMA system depends upon the size of the Hadamard matrix. In the hybrid system, the number of users can be increased indefinitely by simply increasing the number of frequency hops once the CDMA system saturates. Bandwidth should, however, permit this.

3. FH is extremely difficult with the MC-CDMA approach as the number of OFDM frames per hop is very large, basically the size of the Hadamard matrix.

4. The data rate in MC-CDMA system is slow as compared to the hybrid system.

7.4   Analytical Performance in Fading Channels and Simulation in AWGN Channels

This section presents the system performance in an AWGN channel based on analytical results and on simulations. It is noted that the hybrid approach, if it is used, requires a large amount of bandwidth both analytically and during simulations. It is emphasized in this section that the main motivation for the design is higher users.

7.4.1   Comparison of DS-CDMA and DS-CDMA-SFH (DS-SFH) Systems

It is assumed that there are K active users, each with a transmitter-receiver pair. Each user employs a channel encoder. We denote by $b_K(t)$ the modulating sequence of the Kth user, which randomly takes values from the set $\{+1,-1\}$ with equal probability. The coded bit duration is indicated by T and the transmitted power by P.

$$h_K(t) = \sum_{l=1}^{L} h_{Kl}\delta(t - \tau_{Kl})\exp(j\phi_{Kl}) \tag{7.17}$$

For each path, the gain k_{Kl} is Rayleigh distributed, and the delay τ_{Kl} is uniformly distributed over $[0,2\pi]$. We note that the average power for each path and each user is $E[h_{Kl}^2] = \sigma_0/2$. The receiver introduces AWGN $n(t)$ with two-sided PSD N0/2.

7.4.2   Noncoherent Class of Signals (DS-CDMA)

Figures 7.8 and 7.9 illustrate the transmitter and receiver of the DS-CDMA system, respectively. We shall first consider DPSK transmission.

The transmitted signal $x_K(t)$ is

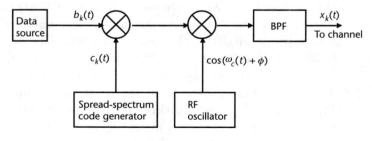

Figure 7.8   DS-CDMA system (transmitter).

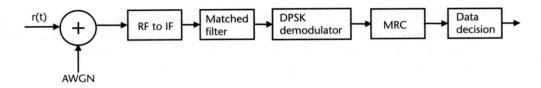

Figure 7.9 DS-CDMA system (receiver).

$$x_K(t) = \mathrm{Re}\left\{\sqrt{2P}b_K(t)c_K(t)\exp\left[j\left(2\pi f_c t + \theta_K\right)\right]\right\} \tag{7.18}$$

where f_c is the carrier frequency, θ_K is the phase introduced by the DPSK modulator, and $c_K(t)$ is a PN sequence with a rate $1/T_c$ and rectangular pulse.

The received signal $r(t)$ is given by

$$r(t) = \sum_{K=1}^{K}\sum_{l=1}^{L}h_{Kl}x_K\left(t - \tau_{Kl}\right)\cos\phi_{Kl} + n(t) \tag{7.19}$$

At the ith receiver, the signal corrupted by noise and interference is filtered, downconverted, and recovered at the output of a filter matched to the spreading code $c_i(t)$. It is sampled at the instants $\lambda T + \tau_{il}$ and is then given to the DPSK demodulator. The demodulated output is denoted by $\Psi_l[\lambda] = \mathrm{Re}\{d_i[\lambda]d_i^*[\lambda-1]\}$ where $d_i[\lambda]$ is the complex notation of the matched filter output. This demodulator output is then given to a maximum ratio combiner (MRC) to yield the final output $\Psi[\lambda] = \dfrac{1}{M}\sum_{l=1}^{M}\Psi_l[\lambda]$, where M is the diversity, $M \leq L$.

The matched filter output $d_i[\lambda]$ comprises

$$d_i[\lambda] = W_i[\lambda] + I_i^{MP}[\lambda] + I_i^{MA}[\lambda] + N_i[\lambda] \tag{7.20}$$

where $W_i[\lambda]$ is the required signal

$$W_i[\lambda] = \sqrt{P/2}\,h_{il}b_i[\lambda]\exp\left(j\phi_{Kl}\right) \tag{7.21}$$

The interference contributions are zero-mean complex Gaussian RVs and comprise I_i^{MP}, which is the multipath interference term caused by $L-1$ other path signals, and I_i^{MA} is the multiaccess interference term caused by $K-1$ other users.

The average SNR is given by the approximation [24] for K users, L paths, and N sequence period:

$$SNR = \gamma = \left[\left(\frac{2E_b}{N_0}\right)^{-1} + \frac{LK-1}{N}\times\frac{1}{3}\right] \tag{7.22}$$

which is the SNR per bit. The constant 1/3 is due to the rectangular chip pulse. This equation is valid for binary DS/SSMA systems. In the case of MRC with diversity of order M, given that the received signal is Rayleigh distributed, the bit error probability P_b at the decoder input is given by [25]

$$P_b = \frac{1}{2^{(2M-1)}(M-1)!(1+\gamma)^M} \sum_{m=0}^{M-1} C_m (M-1+m)! \left(\frac{\gamma}{\gamma+1}\right)^m \tag{7.23}$$

where

$$C_m = \frac{1}{m!} \sum_{n=0}^{M-m-1} \binom{2M-1}{n} \tag{7.24}$$

At the output of the (n, k) block decoder, the probability of bit error is given by [12]

$$P_c = \sum_{j=t+1}^{n} \binom{n}{j} P_b^j (1-P_b)^{n-j} \tag{7.25}$$

where t denotes the error-correction capability of the code.

7.4.3 DS-CDMA-SFH

The transmitter model for DS-SFH is depicted in Figure 7.10. The general derivation for this system is similar, the only difference being that the system uses slow FH. The considered system is asynchronous as this is the more realistic case. Even when synchronism can be achieved between individual user clocks, radio signals will not

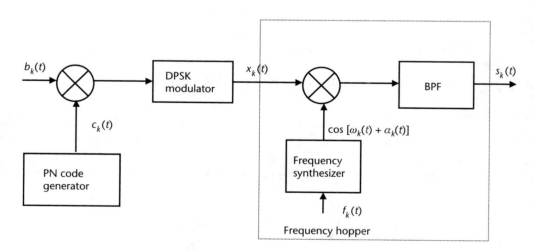

Figure 7.10 Transmitter model for DS-SFH.

arrive synchronously with each user due to propagation delays. The probability of bit error for DPSK is given by (7.23).

However, if two users transmit simultaneously in the same frequency band, a collision, or hit, occurs. In this case, we assume the probability of error as 0.5. Hence, the overall probability of bit error for an FH signal is

$$P_b = Pb(1 - P_{hit}) + \frac{1}{2}P_{hit} \tag{7.26}$$

where P_{hit} is the probability of a hit, derived as discussed next.

If there are q possible hopping channels, there is a $\frac{1}{q}$ probability that a given interfering user will be present in the desired user's slot. Hence, for $K - 1$ interfering users, the probability that at least one is present in the desired frequency slot is

$$P_{hit} = 1 - \left\{ 1 - \frac{1}{q}\left(1 + \frac{1}{N_b}\right) \right\}^{K-1} \tag{7.27}$$

where N_b is the number of bits per hop. We take this value as 1; that is one hop per bit for slow hopping.

At the output of the (n, k) block decoder, the probability of bit error is given by [25]

$$P_e = \sum_{j=t+1}^{n} \binom{n}{j} P_b^j (1 - P_b)^{n-j} \tag{7.28}$$

where t denotes the error-correction capability of the code.

Numerical Results   We now examine the performance of the DS-CDMA and DS-SFH systems for $L = 8$ and $N = 300$.

In Figure 7.11, we compare the DS-CDMA with DS-SFH for various diversity values ranging from one to four. We have assumed the number of hops for the SFH system to be 30 with a processing gain of 10. This is equivalent in terms of bandwidth to a CDMA system with a processing gain of 300. We note that for the same number of users, the SFH system is inferior to CDMA. The asymptotic behavior of both the sets of curves is due to MAI as expected. In both cases, the limiting error will be due to other users. The poor BER of DS-SFH is explained by the fact that there is frequency clashing at the hops, which can be remedied by increasing the bandwidth, reducing the multipath interference, and increasing SNR. However, DS-SFH has its advantages in that the near-far effect is reduced and synchronization is easier (within a fraction of hop time as compared to DS-CDMA, where it has to be within a fraction of chip time). Figure 7.12 compares DS-CDMA and DS-SFH for 10 users using the diversity factor 1.

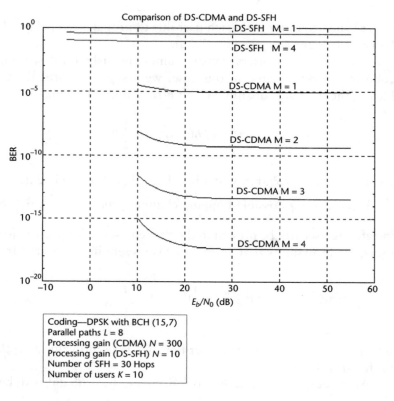

Figure 7.11 Comparison of DS-CDMA and DS-SFH for 10 users (DPSK).

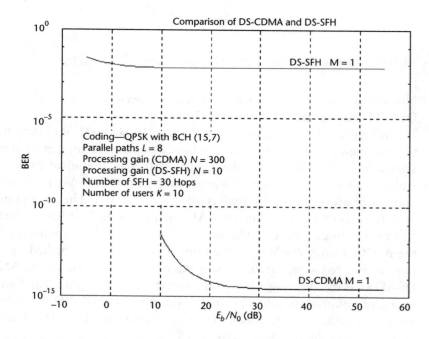

Figure 7.12 Comparison of DS-CDMA and DS-SFH for 10 users (QPSK).

7.4.4   Coherent Class of Signals

In this class of signals, we shall consider QPSK. We use the following equation, with $M = 4$:

The probability of bit error for M- QAM is given by [12]

$$P_b = 4\left(1 - \frac{1}{\sqrt{M}}\right)\frac{1}{2}\, erfc\left(\sqrt{\frac{E_b}{N_0}}\right) \qquad (7.29)$$

The rest of the derivation for both classes of systems is similar.

7.4.5   OFDM-CDMA-SFH (Hybrid)

The output $S_K(t)$ of the OFDM modulator is given by

$$S_K(t) = \sum_m s_{km}(t - mT_B) \qquad (7.30)$$

where each waveform $s_{km}(t)$ is obtained by modulating a block of N consecutive bits $b_k^m[n]$, by n SCs $f_n = n\Delta f, \Delta f = \frac{1}{T_0} \le \frac{1}{NT}, n = 0,\dots N-1$:

$$s_{km}(t) = \sum_{n=0}^{N-1} b_k^m[n]\exp(j2\pi f_n t)g_T(t) \qquad (7.31)$$

where $g_T(t)$ is a rectangular pulse of duration $T_B = NT$. The samples of the OFDM signal $s_{km}(t)$ are generated using IFFT as discussed earlier.

We now frequency-hop the OFDM signal (7.31) according to the kth hopping pattern $f_k(t)$ derived from a hop set of q possibilities. Hence, the transmitted signal can be expressed as

$$S_m(t) = V_m S_k(t)\cos(2\pi f_c t + 2\pi f_k t + \theta_m + \theta_k) \qquad (7.32)$$

where

$$V_m = \sqrt{A_{mc}^2 + A_{ms}^2}\ \text{and}\ \Theta_m = \tan^{-1}\left(\frac{A_{ms}}{A_{mc}}\right)$$

A_{mc} and A_{ms} are the information-bearing signal amplitudes of the quadrature carriers and $S_k(t)$ is the signal pulse. The hop frequency $f_k(t)$ and the phase θ_k are constant over one hop. The phase θ_k represents the phase shift introduced by the frequency hopper when it switches from one frequency to another.

We correct the SNR for the number of parallel paths L, number of users K, and sequence period N.

Due to the GI, an SNR loss is taken into account as SNR $\times \eta_g$. Normally, we assume $\eta_g = 0.8$.

We can now express the dehopper output as $r_i(t)$ at the jth hop of the ith receiver:

$$r_i(t) = V_m \sum_{k=1}^{K} \sum_{l=1}^{L} b_{kl} S_K(t - t_{Kl}) \delta\left[f_k(t - \tau_{ij}), f_i(t - \tau_{ij}) \right] \exp\left(A_{klj} \right) + N(t) \qquad (7.33)$$

where δ is the Dirac delta function and A_{klj} includes the phases introduced by the QAM modulator, hopper, radio channel, and dehopper. The noise contribution $N(t)$ is white complex Gaussian noise.

After sampling at frequency f_s, we have $r_i[m,n] = r_i(mT_B + n/f_s)$. The cyclic prefix is removed, and the resulting block of N_T samples enters the FFT processor:

$$d[m,n] = \sum_{k=1}^{N_T-1} r_i[m, k] \exp\left(-j2\pi nk/N_T\right) \qquad (7.34)$$

Finally, the QAM modulator outputs the decision variables $\alpha[m,n]$.
The probability of bit error for M-QAM is given by [12]

$$P_b = 4\left(1 - \frac{1}{\sqrt{M}}\right) \frac{1}{2} \, erfc\left(\sqrt{\frac{E_b}{N_0}}\right) \qquad (7.35)$$

In our case, $M = 16$. We correct this for probability of hit using (7.26), then correct for error-correction coding using (7.28).

Numerical Results   In Figure 7.13 we now compare DS-CDMA with OFDM-CDMA-SFH (hybrid) with a 16-QAM modulation. Since OFDM has no multipath problem, $L = 1$. We also assume a diversity of $M = 1$. For CDMA, we assume perfect power control and perfect orthogonality between the PN sequences.

We have used BCH (15,7) error correcting code. The asymptotic behavior of the hybrid curve is due to MAI. The CDMA system performs better. At a higher diversity it will perform even better. We also have a cyclic prefix of 20% ($\eta_g = 0.8$) for the OFDM system, as discussed above. However, the hybrid is a multicarrier system. This is not evident from the preceding analytical equations. It also incorporates additional users as discussed below. If we reduce the cyclic prefix further, the SNR improves, but at the cost of poor synchronization. Suppose under normal circumstances our hybrid system cannot tolerate more than 10 users and that users in excess of this number will cause collisions and consequent poor bit-error capability. This is where the advantage of our approach pays off. We can pass the burden of the additional users to the CDMA system. This means that we are looking at a scenario of 10 users operating simultaneously using the hybrid concept, but in actuality there will be many more, the number being limited only by the capability of the respective CDMA systems. For example, if the CDMA system can at the most tolerate 20 users in the uplink due to MAI problems, our hybrid system will actually allow $10 \times 20 = 200$ users. In addition, the near-far effect in the uplink is also taken care of due to FH.

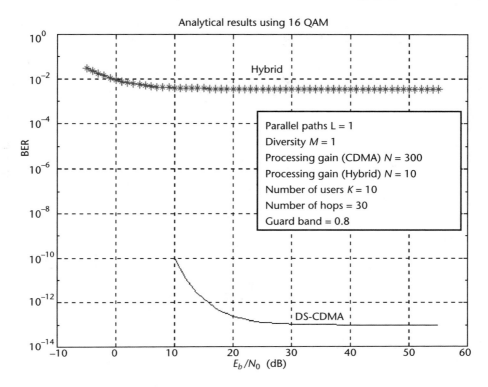

Figure 7.13   Comparison of DS-CDMA with OFDM-CDMA-SFH (16-QAM).

7.4.6   Simulations

The preceding results were analytical. We now examine the simulation behavior of OFDM, OFDM-CDMA, and the hybrid approach in AWGN with 16-QAM modulation. The result is shown in Figure 7.14.

We notice a slight discrepancy between the OFDM-CDMA curve and the plain OFDM curve caused by the spreading and despreading operation. Theoretically, in the presence of AWGN, there should be no discrepancy in the BER curves. Based on the foregoing analytical solutions, we have adjusted the processing gain and the number of hops so that the hybrid system offers a better BER as compared to an OFDM-CDMA system. Specifically, the processing gain was 2 for the OFDM-CDMA system and 16 for the hybrid systems (this low figure was chosen to reduce the program running time), and the number of hops was 4. This makes the hybrid approach perform tolerably as compared to OFDM/CDMA (or MC-CDMA) and plain OFDM modulation. This approach has been followed throughout this book. The idea behind this decision is to ensure that we have a hybrid system tuned to offer an acceptable BER in AWGN. Thereafter, we can investigate its behavior in fading channels because, more importantly (as has been stressed throughout this book), the system offers more users FH capability and its consequent advantages. This is necessary, as we have seen from the analytical results that frequency-hopped systems have higher BERs as compared to

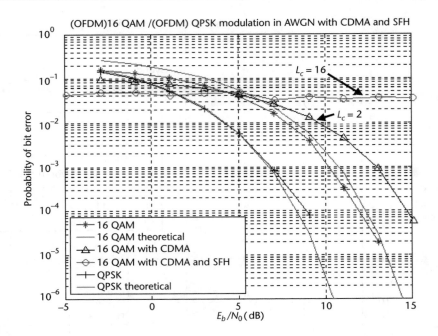

Figure 7.14 16-QAM modulation in AWGN with CDMA and SFH.

nonhopped ones for the same bandwidth [10]. Note that the hybrid curve is level due to saturation caused by multiuser interference. If we increase the chip spread factor or increase the number of hops, the bandwidth increases and the BER improves.

7.4.7 Summary

In this section, we have examined the analytical performance of DS-CDMA, DS-SFH, and OFDM-CDMA-SFH systems in fading channels and simulated their performance in AWGN. We arrived at the following conclusions:

1. We found that the BER of DS-SFH mode is inferior to that of DS-CDMA due to hopping constraints. This study was carried out for DPSK as well as QPSK.
2. We then compared DS-CDMA with OFDM-CDMA-SFH (hybrid). We noted a similar result. The modulation used for comparison was 16-QAM.
3. We next showed that if the hybrid system cannot handle more users due to collision, the CDMA system takes over and improves the number of users multiplicatively.
4. The preceding results required that we tune the hybrid system by adjusting the processing gain and the number of hops to make it viable in terms of BER, the idea being that we could then exploit the other attractive properties of the hybrid system.

7.5 Performance in Fading Channels with Perfect Estimation

This section investigates the performance of hybrid techniques in fading channels using perfect estimation. During this simulation, a Rayleigh channel and a Rician channel are chosen. Once again, the hybrid system places demands upon bandwidth if it is used.

7.5.1 FD Modeling

In this case we shall consider two types of modulation, QPSK and 16-QAM. The system behavior shall be analyzed both in Rayleigh as well as Rician fading environments. We first need to cover certain basic concepts as regards the modus operandi of the simulations. Broadband mobile radio channels can be characterized as linear time-variant filters with a complex equivalent lowpass IR as $h(\tau,t)$ (τ = excess delay; t = time variability). In the FD we use its Fourier transform $H(f;t.)$. However, over a short time interval, the shadowing can be considered constant, and in such a case, we can characterize the channel as a WSSUS. This means that we can apply the theory of linear time-variant filters [25, 26]. This assumption is appropriate for describing the small-scale behavior of the channel, which is suitable and sufficient for our purposes. The nature of the channel will be frequency selective due to the high rates of transmission considered.

We can model the time variability of a narrowband flat fading signal by shaping the spectrum of a complex Gaussian noise process. This is called Jakes's fading model [27]. The resulting colored, complex Gaussian sequence can then be characterized by the spaced-time correlation function $\phi_H(\Delta t)$ and its Fourier transform, the Doppler power spectrum $S_H(\lambda)$ [25].

The idea of FD channel modeling is to shape the spectrum of a complex Gaussian noise process in order to apply the required spaced-frequency correlation function $\phi_H(\Delta f)$. Similarly, the correlation function is specified by its FT, that is, the DPS $\phi_h(\tau)$. In agreement with measurements reported in [28], the shape of the DPS was defined as shown in Figure 7.15 and (7.36).

The DPS is characterized by four parameters: ρ^2 [dB], the power of the direct ray, Π [dB/ns], the power density of the constant-level part, and A [dB/ns], the slope of the exponentially decaying part.

$\gamma = A \cdot \ln(10)/10$ [1/ns] is a decay exponent used for notational convenience.

$$\phi_h(t) = \begin{cases} 0 & \tau < 0 \\ \rho^2\,\delta(\tau) & \tau = 0 \\ \Pi & 0 < \tau \leq \tau_1 \\ \Pi e^{-\gamma(\tau-\tau_1)} & \tau > \tau_1 \end{cases} \qquad (7.36)$$

We can derive analytically from this DPS, other channel parameters like NRP, Rician K-factor, RDS and coherence bandwidth, as well as higher-order statistical parameters, such as the level crossing rate and the average bandwidth of fades [29].

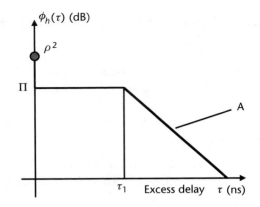

Figure 7.15   Model of the DPS.

The proposed simulation system is shown in Figure 7.16. The amplitude distribution of the TF $H(f)$ is Rayleigh because $H(f)$ is a complex Gaussian noise process. A Rician fading channel may be simulated by adding a phasor given by $h(0) = \rho \cdot e^{j\theta_\rho}$ to $H(f)$, representing the direct path. Otherwise, for a Rayleigh channel $\rho = 0$.

7.5.2   Analytical Evaluation of the BER

The analytical expressions for the BER evaluation are derived in this section. We start our analysis with defining the symbol transmitted on the kth SC as x_{ki}, which is

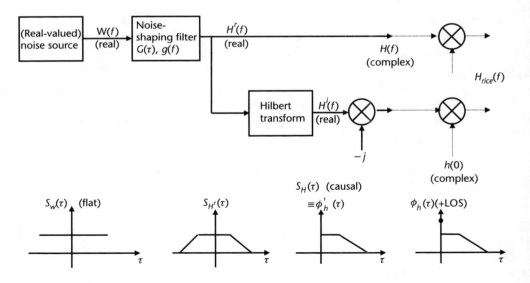

Figure 7.16   FD simulation of the frequency selective radio channel.

an element of the symbol set $\{x_{km}\}$, $m = 1, 2, ..., M$. At the receiver's site, an optimum detector will choose the symbol x_{kj}, which minimizes the distance metric

$$M_d\left(x_{kj}\right) = \left|y_k - \hat{h}_k x_{kj}\right|^2 \tag{7.37}$$

and is thus assumed most likely to be the transmitted symbol. The term $\hat{h}_k x_{kj}$ in this equation accounts for the channel estimation. An error occurs when the metric calculated for a symbol $x_{kj} \neq x_{ki}$ is smaller than the metric for the transmitted symbol x_{ki}. This is written as

$$P_e = \Pr\left\{M_d\left(x_{kj}\right) < M_d\left(x_{ki}\right)\right\} = \Pr\{D > 0\} \tag{7.38}$$

$D = M_d(x_{kj}) - M_d(x_{ki})$ is called the decision variable. D is found

$$D = y_k \hat{h}_k^*\left(x_{ki}^* - x_{kj}^*\right) + y_k^* \hat{h}_k\left(x_{ki} - x_{kj}\right) + \left|\hat{h}_k\right|^2\left(\left|x_{kj}\right|^2 - \left|x_{ki}\right|^2\right) \tag{7.39}$$

where * denotes the complex conjugate. From the channel and system models, y_k is known to be a complex Gaussian RV. The same holds for $\hat{h}_k$ because it is an estimate of the complex Gaussian TF $H_{Rice}(f)$. Thus, the decision variable D is a special case of the generic quadratic form in a complex-valued Gaussian RV. In our case, $L = 1$, considering one transmitted symbol over one (sub)channel.

$$D = \sum_{k=1}^{L}\left[A|X_k|^2 + B|Y_k|^2 + CX_k Y_k^* + C^* X_k^* Y_k\right] \tag{7.40}$$

The probability of error is the probability that $D < 0$, which is evaluated in [25, Appendix B]. This probability is denoted as the integral over the PDF of D

$$P_e = \Pr\{D < 0\} = \int_{-\infty}^{0} p(D)dD \tag{7.41}$$

For $L = 1$, the solution to this integral is written

$$P_e = Q_1(a,b) - \frac{v_2/v_1}{1 + v_2/v_1} I_0(ab)e^{-\frac{1}{2}(a^2 + b^2)} \tag{7.42}$$

where $I_n(x)$ is the nth-order modified Bessel function of the first kind and $Q_1(a,b)$ is the Marcum's Q function, which can be expressed in terms of Bessel functions as

$$Q_1(a,b) = e^{-\frac{1}{2}(a^2 + b^2)} \sum_{n=0}^{\infty}(a/b)^n I_n(ab) \tag{7.43}$$

The parameters a, b, v_1, and v_2 must be related to the moments of X_k and Y_k and to the constants A, B, and C. As given in [25], this is obtained by

$$a = \left[\frac{2v_1^2 v_2 (\alpha_1 v_2 - \alpha_2)}{(v_1 + v_2)^2}\right]^{1/2}$$

$$b = \left[\frac{2v_2 v_2^2 (\alpha_1 v_1 - \alpha_2)}{(v_1 + v_2)^2}\right]^{1/2}$$

$$v_1 = \sqrt{w^2 + \frac{1}{4\left(\mu_{xx}\mu_{yy} - |\mu_{xy}|^2\right)(|C|^2 - AB)}} - w$$

$$v_2 = \sqrt{w^2 + \frac{1}{4\left(\mu_{xx}\mu_{yy} - |\mu_{xy}|^2\right)(|C|^2 - AB)}} + w \qquad (7.44)$$

$$w = \frac{A\mu_{xx} + B\mu_{yy} + C\mu_{xy}^* + C^*\mu_{xy}}{4\left(\mu_{xx}\mu_{yy} - |\mu_{xy}|^2\right)(|C|^2 - AB)}$$

$$\alpha_1 = 2\left(|C|^2 - AB\right)\left(|\overline{X}_1|^2 \mu_{yy} + |\overline{Y}_1|^2 \mu_{xx} - \overline{X}_1^*\overline{Y}_1\mu_{xy} - \overline{X}_1\overline{Y}_1^* - \overline{X}_1\overline{Y}_1^*\mu_{xy}^*\right)$$

$$\alpha_2 = A|\overline{X}_1|^2 + B|\overline{Y}_1|^2 + C\overline{X}_1^*\overline{Y}_1 + C^*\overline{X}_1\overline{Y}_1^*$$

These equations are applied to our problem by comparing (7.39) and (7.40). By letting $Y_1 = y_k$ and $X_1 = \hat{h}_k$ in (7.39), the constants $A = |x_{kj}|^2 - |x_{ki}|^2$ $B = 0$ and $C = x_{ki} - x_{kj}$ are found, representing the properties of the modulation scheme. The behavior of the channel and of the channel-estimation technique are expressed by the first and second moments of the RVs X_1 and Y_1. These are

$$\overline{X}_1 = E\{\hat{h}_k\}$$

$$\overline{Y}_1 = E\{y_k\}$$

$$\mu_{xx} = \frac{1}{2}\left[E\{|\hat{h}_k|^2\} - |\overline{X}_1|^2\right] \qquad (7.45)$$

$$\mu_{yy} = \frac{1}{2}\left[E\{|\hat{y}_k|^2\} - |\overline{Y}_1|^2\right]$$

$$\mu_{xy} = \frac{1}{2}\left[E\{|\hat{h}_k y^* k|^2\} - \overline{X}_1\overline{Y}_1^*\right]$$

7.5.3 Coherent Detection with Perfect Channel Estimation

The signal received at the kth SC is defined as $y_k = x_k + n_k$. Perfect channel estimation means that the receiver has exact knowledge of the attenuation factor h_k, denoted by $\hat{h}_k = h_k$. Considering the transmitted symbol x_{ki} as a constant yields

$$\overline{X}_1 = E\{h_k\} = \rho \times e^{j\theta_p}$$

$$\overline{Y}_1 = E\{x_{ki}h_k + n_k\} = x_{ki}E\{h_k\} + E\{n_k\} = x_{ki}\rho \times e^{j\theta_p}$$

$$\mu_{xx} = \frac{1}{2}\left[E\{|h_k|^2\} - |\overline{X}_1|^2\right] = \frac{1}{2}\left[-nrp - \rho^2\right]$$

$$\mu_{yy} = \frac{1}{2}\left[E\{|x_{ki}h_k + n_k|^2\} - |\overline{Y}_1|^2\right] = \frac{1}{2}\left[|x_{ki}|^2(NRP - \rho^2) + N_0\right]$$

$$\mu_{xy} = \frac{1}{2}\left[E\{h_k(x_{ki}h_k + n_k)^*\} - \overline{X}_1\overline{Y}_1^*\right] = \frac{1}{2}x^*_{ki}[NRP - \rho^2]$$

(7.46)

where $NRP = E\{|h_k|^2\}$ is the normalized received power that can be expressed in terms of the parameters of the FD channel model $NPR = \rho^2 = \prod(\tau_1 = 1/\gamma)$.

7.5.4   Calculation of the Parameters

In QPSK the transmitted symbols are taken from the set $x_{km} = \{1, j, -1, -j\}$. Note that $E\{|x_{km}|^2\} = 1$. Thus, the transmitted energy per symbol is unity. QPSK transmits two bits per symbol on an orthogonal basis. Thus, the error probabilities can be analyzed independently, and the BER equals their average. Suitable parameters for A and C are found by assigning, for example, $x_{ki} = 1$ and $x_{kj} = \{j, -j\}$.

In order to calculate the probability that x_{kj} has been detected when x_{ki} was transmitted, we divide the I/Q plane into two parts. An error will occur when the received symbol falls within the half-plane that is closer to x_{kj} than to x_{ki}. Furthermore, it will be noticed from Figure 7.17 that it is not necessary to evaluate the two-error event explicitly. The overlapping one-error events account for one error each in this region.

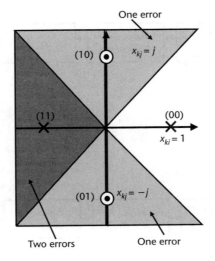

Figure 7.17   QPSK.

The 16-QAM class of signals can be evaluated without any errors. We need to consider 4 different transmitted symbols occurring with equal probability and 24 error events, some with a negative sign. Table 7.3 lists the symbols x_{ki} and x_{kj} to be used. The complex signal constellations $x_k = a + jb$ are denoted (a,b). Error events that must be subtracted in the final result are denoted $(a,b)^{-1}$.

7.5.5 Simulations with Perfect Channel Estimation

We shall first examine the behavior of the proposed system assuming perfect channel estimation and perfect synchronization. Later in this book, we shall examine the implications of realistic channel estimation. The latter is dependent upon the type of estimation technique used.

We shall simulate the OFDM system in two different models, Rayleigh and Rician, discussed later. We now discuss the OFDM system simulation model. For the simulation of the OFDM system, we shall use the following model as given in Figure 7.18.

Table 7.3 Transmitted Symbols and Error Events for the Evaluation of 16-QAM Modulation*

Transmitted Symbol x_{ki}	Error Symbols x_{kj}
$(-3,3)$	$(-1,3),(3,3),(7,3)^{-1}, (-3,1),(-3,-3),(-3,-7)^{-1}$
$(-1,3)$	$(-3,3),(1,3),(5,3), (-1,1),(-1,-3),(-1,-7)^{-1}$
$(-3,1)$	$(-1,1),(3,1),(7,1)^{-1}, (-3,3),(-3,-1),(-3,-5)$
$(-1,1)$	$(-3,1),(1,1),(5,1), (-1,3),(-1,-1),(-1,-5)$

* All values must be divided by $\sqrt{10}$ to have an average power of one.

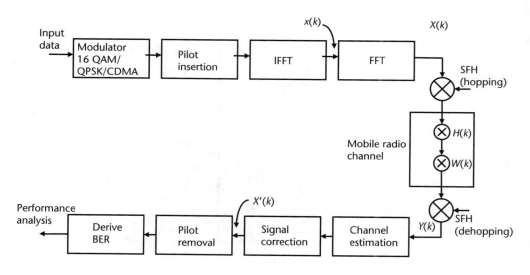

Figure 7.18 OFDM system simulation model.

We are not considering ISI and ICI multipath effects in these simulations for the sake of simplicity. We are also assuming perfect synchronization and are not employing any coding. In Figure 7.18, the binary input data is modulated with 16-QAM/QPSK modulation, with or without CDMA. In the presence of CDMA, we shall assume a chip spread factor of two. In the hybrid cases we shall assume a chip spread factor of 16. We would like to have a higher spreading factor, but this results in unduly long simulations. However, this does affect our conclusions, as we shall see. The number of hops in cases of SFH simulation is taken as four, representing four users. Since channel estimation is being considered in this book, we plan to carry out this estimation using pilot signals. The pilot signals are inserted between the modulated signals. This aspect shall be extensively discussed in Section 7.6.2. IFFT is then performed on this signal in order to obtain an OFDM signal with pilots. Since we are planning to use an FD model of the channel as discussed earlier, the OFDM signal is transformed into FD by performing an FFT. By this FFT, the OFDM signal in TD ($x(k)$) is transformed into FD ($X(k)$). The signal is then slow frequency-hopped. In the FD, we can just multiply the OFDM signal with the channel TF $H(k)$. In the mobile radio channel, the OFDM signal is contaminated with multipath effects, and on this the AWGN noise ($W(k)$) is superimposed. This results in the received OFDM signal $Y(k)$. The signal is then dehopped on reception. Using this OFDM signal with its pilot signals, the channel TF is estimated using the TDP method, also discussed Section 7.6.2. Presently, in this section, we assume perfect estimation. The contaminated OFDM signal is now presumed to have been corrected. This aspect of signal correction is also discussed Section 7.6.2. Presently, since we have assumed perfect estimation, there is no need for this correction, and we proceed to remove the pilot signals retaining only the information SCs and then derive the BER of the system.

7.5.5.1   Types of Channels Considered for Simulation

In our proposed analysis we shall consider two types of channels, one Rayleigh and one Rician [30]. Table 7.4 lists the channel model parameters. Table 7.5 lists the channel parameters for the corresponding channels (used in analytical runs). The

Table 7.4   Channel Model Parameters

Channel	LOS [dB]	A [dB]	τ_1 [ns]	π [dB/ns]	τ_{max} [ns]
Rayleigh	−12.7	0.12	60	−20.1	300
Rician	−0.6	0.17	0	−23	150

Table 7.5   Channel Parameters

Channel	NRP [dB]	τ_{rms} [ns]	K-factor [dB]
Rayleigh	0	42.1	−12.4
Rician	0	12.5	8.3

corresponding channel IR, TF, PDF, and CDF for both these channels are shown in Figures 7.19 and 7.20.

Figure 7.19(c) shows that this channel is a Rayleigh fading indoor radio channel as it fits the Rayleigh distribution. The maximum delay time of the reflected waves is around 300 ns.

We can see from Figure 7.20(c) that this channel is a Rician channel since it does not fit the Rayleigh distribution. Hence, this is a Rician fading indoor radio channel. The maximum delay of the indirect waves of an impulse in this channel is around 150 ns.

We now analyze the behavior of 16-QAM and QPSK signals with and without SFH and with and without CDMA in the presence of perfect channel estimation and

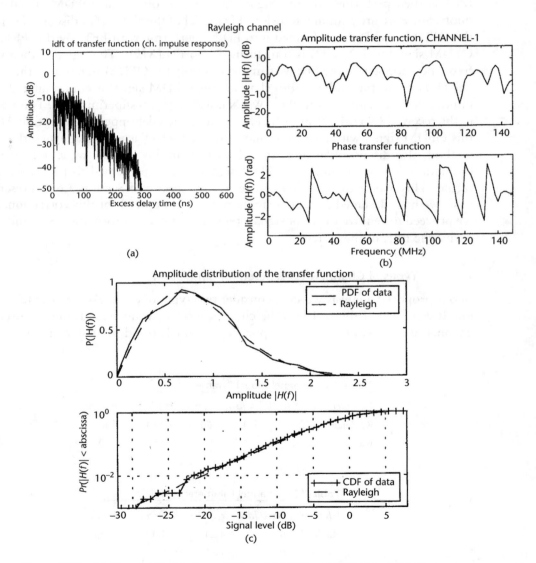

Figure 7.19   (a) Channel IR, (b) channel TF, and (c) PDF and CDF of Rayleigh channel.

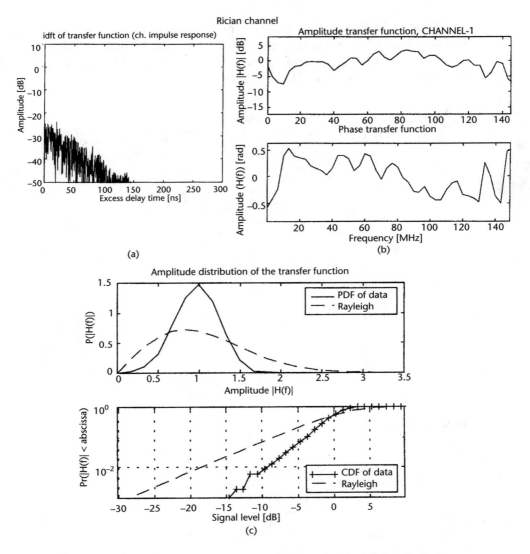

Figure 7.20 (a) Channel IR, (b) channel TF, and (c) PDF and CDF of Rician channel.

perfect synchronization. We are doing this for the case of a single user with a chip spread factor of two. The number of hops and the bandwidth have been chosen so as to yield very low bit rates. These aspects have been discussed earlier. If the number of users is increased, then the BER will deteriorate due to frequency clashing at the hops or due to MAI at the CDMA end. We hope that because we are using a DS-CDMA system combined with OFDM, we can achieve better synchronization in the uplink than has hitherto been possible with pure DS-CDMA systems. In such a case, the effects of MAI will not be so severe and the deterioration of BER will occur mostly due to frequency collision at the hops. This can be remedied by adjusting the bandwidth for the number of users.

Figure 7.21 shows the simulation result and the theoretical result for a QPSK transmission. We note that the analytical result exactly matches the simulated run. We now compare this performance with QPSK with CDMA and QPSK with CDMA and SFH. This is shown in Figure 7.22 for a Rayleigh channel.

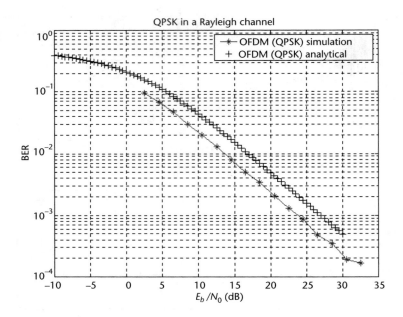

Figure 7.21 QPSK in Rayleigh fading channel.

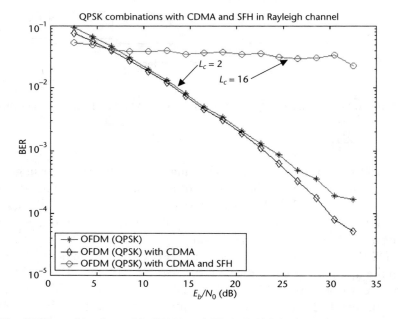

Figure 7.22 QPSK combinations with CDMA and SFH in Rayleigh channel.

We note that QPSK with CDMA and SFH does not perform well. It requires a lot of bandwidth. We now examine the performance of 16-QAM modulation in such a channel. The constellation is shown in Figure 7.23.

We now compare the simulated and analytical result for 16-QAM in a Rayleigh channel. This is shown in Figure 7.24. We note that the analytical result exactly

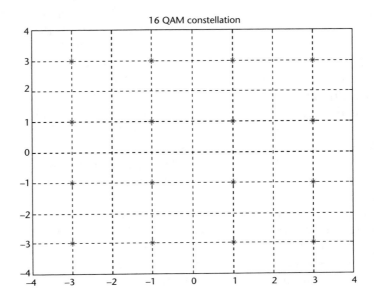

Figure 7.23 16-QAM constellation.

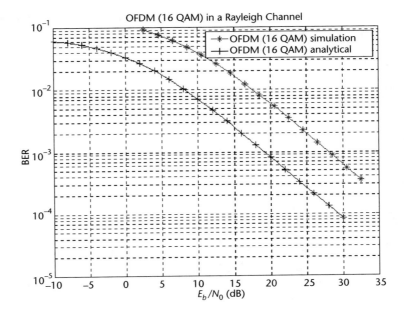

Figure 7.24 16-QAM in a Rayleigh channel.

matches the simulated run. We now compare this performance with 16-QAM with CDMA and 16-QAM with CDMA and SFH. This is shown in Figure 7.25 for a Rayleigh channel.

Once again the BER of the hybrid system is poor. If we need to improve the BER of the hybrid system, we need to increase the bandwidth even more. We now examine the overall performance of both these signal schemes in a Rayleigh channel. This is shown in Figure 7.26, which summarizes our conclusion that in a Rayleigh fading channel, 16-QAM with CDMA and SFH will perform the best if adequate bandwidth is available.

The overall result for the Rician channel is given in Figures 7.27, 7.28, and 7.29. We note that once again 16-QAM with CDMA and SFH will require adequate bandwidth for viable bit rates.

Finally, we examine the behavior both classes of signals in both types of channels for a fair comparison. We choose QPSK and 16-QAM without CDMA and without SFH (see Figure 7.30).

We note that QPSK is the superior modulation in both Rayleigh and Rician fading channels. If we need to use the hybrid approach, we need to compromise on BER.

Note that SFH has not been proposed for improving BER, but to combat the near-far effect of CDMA. Similarly, CDMA has been proposed not for BER improvement, but to increase the number of users in the presence of SFH. We now examine the behavior of this approach in the presence of realistic channel estimation. We just need to determine which class of signals is superior in the presence of realistic channel estimation.

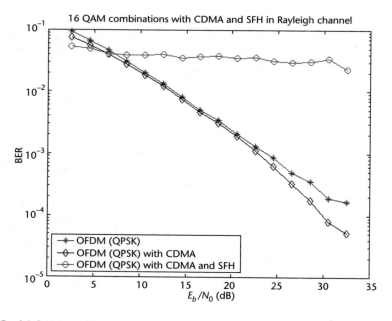

Figure 7.25 16-QAM combinations with CDMA and SFH in a Rayleigh channel.

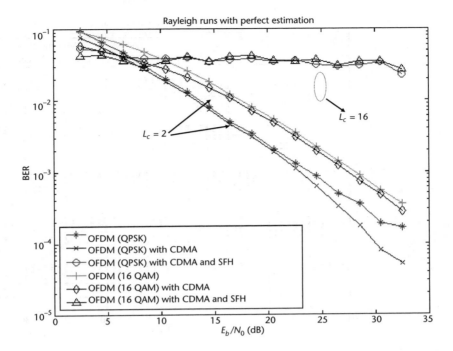

Figure 7.26   Rayleigh runs with perfect estimation.

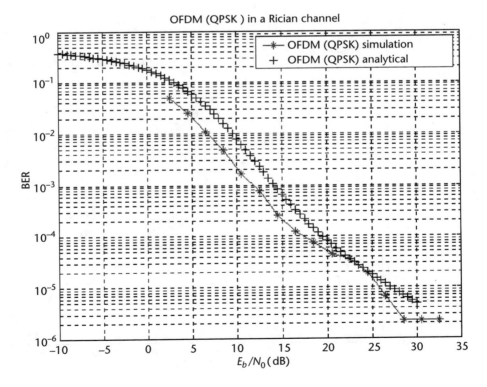

Figure 7.27   QPSK in a Rician channel.

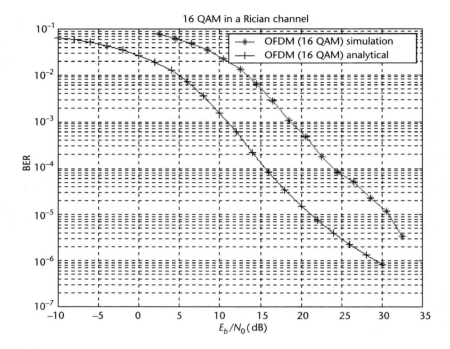

Figure 7.28   16-QAM in a Rician channel.

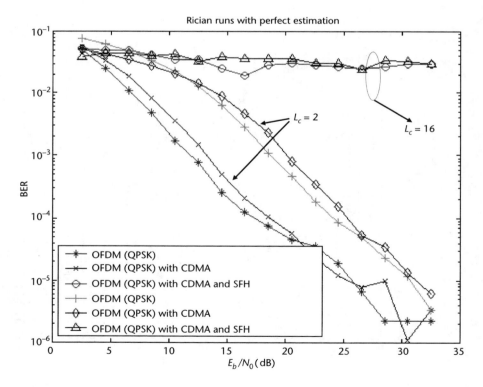

Figure 7.29   Rician runs with perfect estimation.

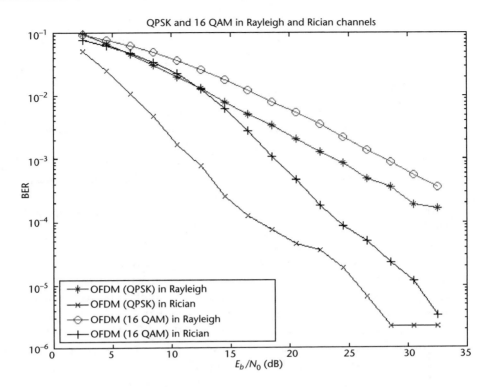

Figure 7.30   QPSK and 16-QAM in Rayleigh and Rician fading channels.

7.5.6   Summary

In this section, the channel was modeled in the FD. Monte Carlo simulations were then carried out based on two classes of signals, QPSK and 16-QAM and assuming perfect channel estimation and perfect synchronization. We arrived at the following conclusions:

1. Using QPSK modulation we found that in both Rayleigh and Rician channels, QPSK performance is superior.
2. We then compared both QPSK and 16-QAM using progressively more complex modulation schemes. We found that in each case QPSK was superior.
3. We know that 16-QAM is the better type of modulation from the point of view of higher throughput. But how does 16-QAM perform in the presence of realistic estimation? This aspect will be investigated in the next section using realistic channel estimation.

7.6   Performance in Fading Channels with Realistic Estimation

This section analyzes the behavior in a fading channel using realistic estimation. The simulations show that QPSK performs better, as it is more robust compared to 16-QAM. The hybrid system once again requires higher bandwidths. However, if

BER is the predominant consideration, then the proposed system can be used by discarding the SFH part.

7.6.1 Baseband Model

This aspect of the book is based on the work done by Tutucu [30]. A baseband model of an OFDM system using pilot-based signal correction is shown in Figure 7.31.

Binary information data is encoded in multiphase signals (QPSK/16-QAM). Then the pilot signals are uniformly inserted into the information data sequence. The function of the IFFT block is to convert the data sequence on length N into N parallel data, modulate it on N SCs, and sum them. After lowpass filtering, the modulated carrier is then transmitted. At the receiver, the pilot-based signal correction is performed after FFT, followed by decoding.

The total N SCs of the OFDM system are arranged as follows. Adjacent L ($L < N$) are grouped together, without overlapping between adjacent groups. In each group, the first SC is used to transmit the pilot signal and thus is called the pilot SC. The rest of the SCs bear information data and thus are called information SCs. Therefore, there are a total of $M = M/L$ pilot SCs and $N - M$ information SCs.

In the following formulas, $n \in [0, N-1]$ denotes the index in discrete TD and $k \in [0, N-1]$ denotes the index in discrete FD. k is further expressed as $k = mL + 1$, with integers $l \in [0, L-1]$ and $m \in [0, M-1]$, where

N = Total number of SCs of the OFDM system;
M = Number of pilot SCs;
L = Number of SCs that each group contains.

Assuming that all of the pilot signals have an equal complex value c, then the OFDM signal modulated on the kth SC can be expressed as

$$X(k) = X(mL + l) = \begin{cases} c, 1 = 0 \\ data, 1 = 1, ..., L - 1 \end{cases} \tag{7.47}$$

The corresponding TD signal $x(n)$ is obtained by IFFT.

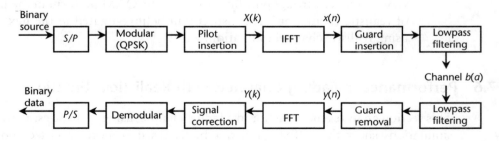

Figure 7.31 Baseband model of the OFDM system with pilot-based signal correction.

The received TD OFDM signal $y(n)$, is a function of the transmitted signal, the channel TF, and AWGN $w(n)$. It can be expressed as

$$y(n) = x(n) \otimes h(n) + w(n), \quad 0 \leq n \leq N - 1 \tag{7.48}$$

where "$\otimes$" denotes convolution.

The received FD signal $Y(k)$ is the FT of $y(n)$, which can be expressed as

$$Y(k) = X(k) \cdot H(k) + I(k) + W(k) \tag{7.49}$$

where $W(k)$ is the FT of $w(n)$. $H(k)$ is recognized as accurate channel TF at the kth SC, which is independent of the transmitted signals $X(k)$ and given by [2]

$$H(k) = \sum_{i=0}^{i-1} h_i e^{j\pi f_{Di} T} \frac{\sin(\pi f_{Di} T)}{\pi f_{Di} T} e^{-j\frac{2\pi \tau_i}{N} k} \tag{7.50}$$

$I(k)$ in (7.49) is the ICI component in the received signal at the kth SC, depending upon the signal values $X(k)$ modulated on all of the other SCs, which is given by

$$I(k) = \frac{1}{N} \sum_{i=0}^{r-1} \sum_{k=0}^{N-1} h_i X(K) \frac{1 - e^{j2\pi(f_{Di} T - k + K)}}{1 - e^{-j\frac{2\pi}{N}(f_{Di} T - k + K)}} e^{-j\frac{2\pi \tau_i}{N} k} \tag{7.51}$$

where h_i is the complex IR of the ith path, f_{Di} is the ith path Doppler frequency shift which causes ICI of the received signals, and τ_i is the ith path delay time normalized by sampling time.

7.6.2   Channel Estimation with TDP Method

This chapter is not on channel-estimation techniques. We are merely selecting one of the many available techniques for channel estimation to enable us to analyze qualitatively the system behavior during realistic channel estimation. This method [31] enables OFDM systems to work in mobile communications where the channels are time variant and frequency selective because no knowledge of the channel TF for the previous OFDM data blocks is required.

This method provides a more accurate estimate of the channel TF, as compared to the normal interpolation method. The lowpass filtering in the transform domain reduces AWGN and ICI significantly. The high-resolution interpolation approach suits well the mathematical model of the channel. The working principle is described below.

The mobile radio channel is time variant. Therefore, the TF for the present data block should be obtained independently of the previous blocks. The SC arrangement in (7.47) is adopted. The proposed channel-estimation method based on pilot signals and transform-domain processing is depicted in Figure 7.32. By L:1 downsampling the received sequence $Y_N(k)$, the samples at the pilot SCs are picked up and a sequence of length M is obtained. Normalizing this sequence to the pilot signal value c gives the noisy channel TF $\dot{H}_M(m)$; that is,

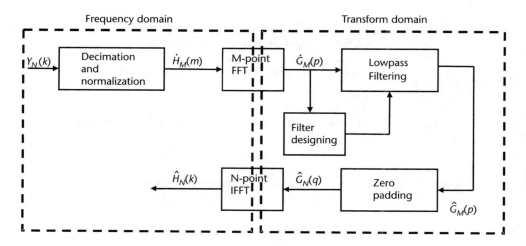

Figure 7.32 Channel-estimation approach with transform domain processing.

$$\dot{H}_M(m) = H_M(m) + \left[I_M(m) + W_M(m)\right]/c, \quad m = 0, \ldots, M-1 \qquad (7.52)$$

Since the AWGN in TD is a zero-mean random process, it can be derived that the noise component in (7.52) is also zero mean and Gaussian distributed [31].

Considering that the channel parameters are unknown and changing from time to time, it is difficult to reduce the noise component in FD. On the contrary, a low-pass filtering in the transform domain is feasible.

The transform domain is defined such that any sequence in this domain is the DFT of its counterpart in the FD. Therefore, a sequence in the transform domain is the "spectral sequence" of its counterpart in the FD. The argument p of the transform domain can be viewed as the "frequency," which reflects the variation speed of an FD function.

The transform-domain representation of $\dot{H}_M(m)$ is then

$$\dot{G}_m(p) = \sum_{m-0}^{M-1} \dot{H}_M(m) \exp\left(-j\frac{2\pi}{M}mp\right) \qquad (7.53)$$

where p is the transform-domain index and $p \in [0, M-1]$. As expected, the signal component in $\dot{G}_m(p)$ is located at the lower "frequency" (around $p = 0$ and $p = M - 1$) region, while the noise component is spread out over the whole frequency region ($p = 0, \ldots, M - 1$).

The lowpass filtering can be realized by simply setting the samples in the "high frequency" region to zero; that is,

$$\hat{G}_M(p) = \begin{cases} \dot{G}_M(p), 0 \le p \le p_c, M - p_c \le p \le M - 1 \\ 0, \text{ otherwise} \end{cases} \qquad (7.54)$$

where p_c is the "cutoff frequency" of the filter in the transform domain. After the filtering, the noise component is reduced to $2P_c/M$ of its original value.

The cutoff frequency p_c of the transform-domain lowpass filter is an important parameter that effects the accuracy of the channel estimation. Its value changes continuously due to the variation of the mobile radio channel. Therefore, an approach is needed to select p_c by dynamically tracking the received signals.

If we look to the energy distribution in the transform domain with respect to different values of p, we see that most of the energy is concentrated at the "low frequency" region where the desired components are located. Therefore, p_c is determined from the following relation:

$$\frac{\sum_{p=0}^{p_c}\left|\overline{G}_M(p)\right|^2 + \sum_{p=M-p_c}^{M-1}\left|\overline{G}_M(p)\right|^2}{\sum_{p=0}^{M-1}\left|\overline{G}_M(p)\right|^2} = R \qquad (7.55)$$

where the numerator is the energy in the "passband," the denominator is the total energy, R is a value between 0.9 and 0.95, and $\overline{G}_M(p)$ is the average of $\dot{G}_M(p)$ of the present data block and 10 previous data blocks.

The simple algorithm for searching for the cutoff frequency p_c moves from the "low-frequency" toward the "high-frequency" region. If the accumulated energy in the low-frequency region is larger than R (a certain percentage of the total energy), the corresponding p is taken as the cutoff frequency p_c. This tracking procedure is shown in Figure 7.32 as "filter designing."

After filtering the $\dot{G}_M(p)$ in the transform domain, which reduces the AWGN significantly, a high-resolution interpolation with zero-padding is performed on $\hat{G}_M(p)$, which gives $\hat{G}_N(q)$ (see Figure 7.32). This is done as follows. First, the M-sample transform-domain sequence $\hat{G}_M(p)$ is extended to an N-sample sequence $\hat{G}_N(q)$ by padding with $N-M$ zero samples at the high-frequency region around $p = M/2$, resulting in

$$\hat{G}_N(q) = \begin{cases} \hat{G}_M(q), 0 \leq q \leq p_c \\ 0, p_c \leq q \leq N - p_c \\ \hat{G}_M(q - N + M), N - p_c \leq q \leq N - 1 \end{cases} \qquad (7.56)$$

This N-sample sequence $\hat{G}_N(q)$, in its physical meaning, is the FT of the desired estimate of the channel TF. By performing an N-point IFFT, the estimated TF with lower noise levels at all SCs is obtained as

$$\hat{H}(k) = a \cdot \sum_{q=0}^{N-1} \hat{G}_N(q) \exp\left(-j\frac{2\pi}{N}qk\right), 0 \leq k \leq N-1 \qquad (7.57)$$

Since the M-point FFT and N-point IFFT are performed between the FD and the transform domain, a constant a is needed for calibration.

A word on the pilot signals: The OFDM system is simulated for two different multipath fading indoor radio channels, one Rayleigh and one Rician. The pilot share rate is 12.5%. The pattern for a pilot share rate of 25% is shown in Figure 7.33 as an example.

By $L = 4$ in the figure, we mean that when we have a pilot signal at SC number x, then we will have another pilot signal at SC number $x + L$. The signal correction technique is shown in Figure 7.34.

In Figure 7.34, the OFDM signal $(X(k))$, in the FD with pilot signals, is sent through the mobile channel with TF $H(k)$. Then the noise is added on top of it, which results in the received signal $Y(k)$. From this received signal, the estimated channel TF $\hat{H}(k)$ is derived using the TDP estimation technique. The conjugate of the estimated channel TF is multiplied by the received signal, which gives us the

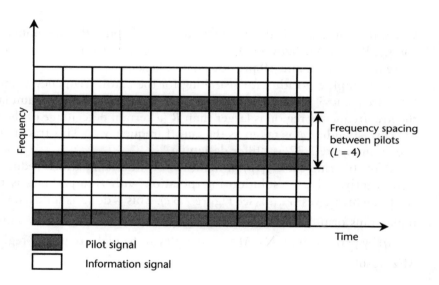

Figure 7.33   Pilot signal pattern for a share rate of 25%.

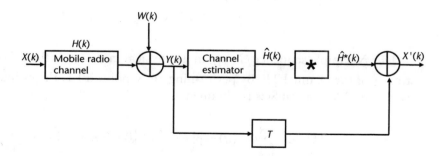

Figure 7.34   OFDM signal correction with the estimated channel TF.

modified transmitted signal $X'(k)$. This signal is then further processed and the BER of the system is derived.

During the design of the OFDM system, we need to take care that the GI is equal to or greater than the maximum delay of the channel in order to get rid of ISI. The Rayleigh channel has a maximum delay time τ_{max} of 300 ns. This is more than that of the Rician channel, which is 150 ns. Hence, both the channel requirements are satisfied.

7.6.3 Simulated OFDM System Parameters

Choosing the GI equal to 300 ns gives us the remaining OFDM parameters

$$T_{guard} = \tau_{max} = 0.3 \, \mu s$$
$$T_S = 12.T_{guard} = 3.6 \, \mu s$$
$$T = T_{guard} + T_S = 3.9 \, \mu s$$
$$\text{Symbol rate} = 1/T = 256.410 \, \text{ksymbols/sec}$$

The minimum data bit rate of our OFDM system has been considered to be 155 Mbps. We therefore calculate the minimum number of information SCs (N_{inf_min}) needed to achieve a data rate of 155 Mbps.

$$R = 2 \times R_{symb} \times N_{inf} \Rightarrow 155,000,000 = 2 \times 256.410 \times N_{inf} \Rightarrow N_{inf}(\min) = 302$$

We shall need roughly twice this number for coding, pilot signals, and the like.

$$\Rightarrow \text{we choose N\_tot (total number of SCs)} = 2 \times N_{inf}(\min) \approx 512$$

Hence, we need a 512-point IFFT and FFT to implement this system. We have arbitrarily chosen the factor of 12 to calculate the symbol interval T_s. Choosing a higher factor will give us a better OFDM performance because we will then have more SCs with which to achieve a 155-Mbps data rate and, consequently, a greater number of SCs for pilot signals, making for better estimation. But we cannot have a very large factor because this will cause the step interval T to be too great, resulting in a lower data rate. Hence, we need to compromise.

We see from Table 7.6 that the pilot share rate of 12.5%, $100 - 12.5 = 87.5\%$ of the transmission rate is used for the data signals and 12.5% for pilot signals. Furthermore, we see that using a pilot share rate of 12.5%, we have a data rate of around 230 Mbps. This value may be decreased to around 155 Mbps by using coding and the like. It is double this value if we use 16-QAM. In our simulations, coding is not considered. We are also assuming perfect synchronization and no ISI and ICI.

7.6.4 Simulation Results for QPSK/16-QAM

We now compare the performance of 16-QAM and QPSK in realistic estimation of Rayleigh and Rician channels in Figure 7.35 and 7.36, repectively.

Table 7.6 Simulated OFDM System Parameters

Parameter	Value (L = 8, 12.5%)	Value (L = 8, 12.5%)
Modulation type	QPSK	16-QAM
Pilot spacing (L)	8	8
Number of SCs (N_tot)	512	512
Number of Pilot SCs	64	64
Minimum number of information carriers	302	302
Number of information SCs (N_inf)	448	448
Number of bits per OFDM symbol	1,024	2,048
Number of data bits per OFDM symbol	896	1,792
Number of pilots per OFDM symbol	128	256
Step interval ($T = T_s + T_G$)	3.9 μs	3.9 μs
Symbol interval (T_s)	3.6 μs	3.6 μs
Guard interval (T_G)	0.3 μs	0.3 μs
Symbol rate (1/T)	256.410 Ksymbols/sec	256.41 Ksymbols/sec
Carrier-frequency spacing ($\Delta f = 1/T_s$)	277.778 kHz	277.778 kHz
Total bandwidth ($f \cdot N$)	142.222 MHz	142.222 MHz
Data rate (R_{data})	229.743 Mbps	459.486 Mbps
Transmission rate (R)	262.564 Mbps	525.128 Mbps

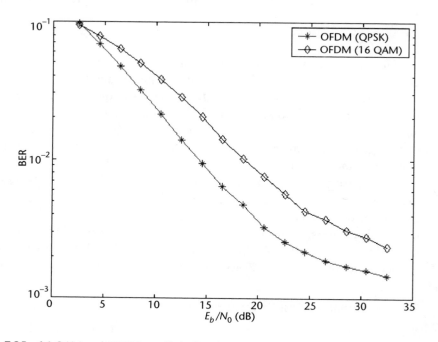

Figure 7.35 16-QAM and QPSK in realistically estimated Rayleigh channel.

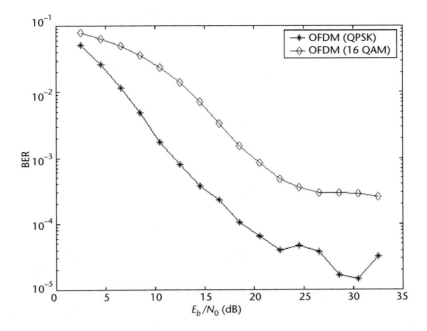

Figure 7.36   16-QAM and QPSK in realistically estimated Rician channel.

We note that in both cases 16-QAM is inferior to QPSK because QPSK belongs to a more robust class of signals. However, the data rate is half that of 16-QAM. It will be noticed that the results are similar to the perfect estimation cases. This implies that our estimator is working perfectly. We now examine the hybrid case with a chip spread of 32. This is shown for both cases, Rayleigh and Rician, in Figures 7.37 and 7.38.

We note that 16-QAM with CDMA and SFH requires a substantial increase in bandwidth, as compared to QPSK or 16-QAM in a Rayleigh channel as well as in a Rician channel. However, this is the price we need to pay if we want to adopt SFH with a view to increasing the number of users or combating the near-far effect.

7.6.5   Summary

In this section we investigated the behavior of QPSK and 16-QAM class of signals in the presence of realistic channel estimation for both Rayleigh and Rician channels. Toward this end, we were required to adopt an estimation method. We adopted the TDP method, based on pilot signal estimation. The TDP method was chosen because it enables OFDM systems to work in mobile communications where the channels are time variant and frequency selective because no knowledge of the channel TF for the previous OFDM data blocks is required. The entire OFDM system parameters were so chosen as to be in conformity with our stated requirement of a minimum bit rate of 155 Mbps in a Rician (60 GHz) channel. A Rayleigh channel was also considered for completeness. We arrived at the following conclusions:

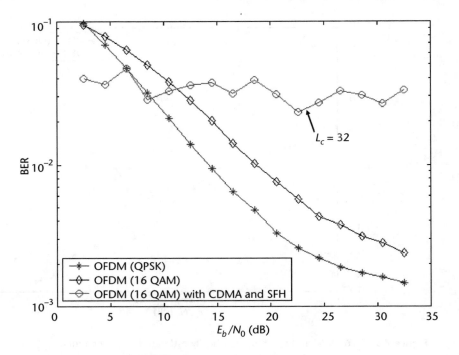

Figure 7.37   16-QAM and QPSK in realistically estimated Rayleigh channel.

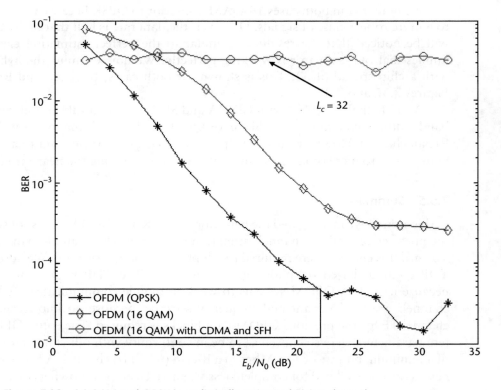

Figure 7.38   16-QAM and QPSK in realistically estimated Rician channel.

1. In order to enable high-bit-rate transmission, it becomes imperative to use 16-QAM in a hybrid approach.

2. Using the hybrid approach places demands upon the bandwidth requirement. We need to exercise this option, however, if we require more users or if we need to combat the near-far effect. Otherwise, we can use the OFDM-CDMA part only, as suggested in this book. It thereby provides bit rates on the order of 525 Mbps conducive with multimedia requirements and at low BERs. The hybrid approach, however, provides for more users. High numbers of users will become a fact in videotelephone systems.

7.7 Conclusions

Multicarrier systems have generated increased interest over the last years, fuelled by a large demand for frequency allocation resulting in a crowded spectrum, as well as a large number of users requiring simultaneous access. This book proposes a new technique called the hybrid approach. This involves the amalgamation of OFDM/CDMA/SFH with a view to improving the number of users and yielding high bit rates based on 16-QAM modulation. It has essentially been developed for the 60-GHz frequency, but it is equally applicable at any other frequency, provided the necessary bandwidth is available.

This proposal pertains to the downlink as well as to the uplink, the only difference being that for the synchronous downlink we can apply orthogonal Walsh-Hadamard sequences leading to the well-known user separation for MC-CDMA systems. On the other hand, in the asynchronous uplink scenario, pseudonoise sequences are used with the drawback of high MAI.

This research has presented and discussed the performance of such a system in AWGN and fading channels, assuming perfect estimation and realistic estimation. We have concluded that the hybrid approach using 16-QAM provides high bit rates with an acceptable BER performance and have proposed a new synchronization scheme for such a system. This scheme is based on an algorithm. In our case, this algorithm is based on fuzzy logic. Simulations have been performed using this algorithm in AWGN as well as a Rician channel and have proved that this technique is excellent for tracking both timing and frequency offsets.

References

[1] Jankiraman, M., and R. Prasad, "Hybrid CDMA/OFDM/SFH: A Novel Solution for Wideband Multimedia Communications," *ACTS Summit,* Sorento, Italy, 1999.

[2] Jankiraman, M., and R. Prasad, "A Novel Algorithmic Synchronization Technique for OFDM Based Wireless Multimedia Communications," *ICC'99,* Vancouver, Canada.

[3] Jankiraman, M., and R. Prasad, "Hybrid CDMA/OFDM/SFH: A Novel Solution for Wideband Multimedia Communications," *MC-SS'99,* Oberpffaffenhausen, Germany, 1999.

[4] Jankiraman, M., and R. Prasad, "A Novel Solution to Wireless Multimedia Application: The Hybrid OFDM/CDMA/SFH Approach," *PIMRC 2000,* London, England, 2000.

[5]   Jankiraman, M., and R. Prasad, "Algorithm Assisted Synchronization of OFDM Systems for Fading Channels," *5th International OFDM Workshop*, Hamburg, Germany, September 2000.

[6]   Jankiraman, M., and R. Prasad, "Performance Evaluation of Hybrid OFDM/CDMA/SFH for Wireless Multimedia," *VTC 2000,* Boston, MA, 2000, pp. 934–941.

[7]   Jankiraman, M., and R. Prasad, "Wireless Multimedia Application: The Hybrid OFDM/CDMA/SFH Approach," *ISSSTA 2000,* Piscataway, NJ, 2000, pp. 387–393.

[8]   Nathan, Yee, and J.-P. Linnartz, *Multi-Carrier Code Division Multiple Access (MC-CDMA): A New Spreading Technique for Communication over Multipath Channels,* Final Report 1993–1994 for MICRO Project 93-101.

[9]   Fazel, K., and L. Papke, "On the Performance of Convolutionally Coded CDMA/OFDM for Mobile Communication System," *Proc. of IEEE PIMRC'93,* September 1993, pp. 468–472.

[10] Prasad, R., *CDMA for Wireless Personal Communications,* Norwood, MA: Artech House, 1996.

[11] Hara, S., and R. Prasad, "Overview of Multicarrier CDMA," *IEEE Communication Mag.,* Vol. 35, No. 12, December 1997, pp. 126–133.

[12] Rappaport, T. S., *Wireless Communications: Principles and Practice,* Upper Saddle River, NJ: Prentice Hall, 1996.

[13] van Nee, R., and R. Prasad, *OFDM for Wireless Multimedia Communications,* Norwood, MA: Artech House, 2000.

[14] Wahlqvist, Mattias, et al., *A Conceptual Study of OFDM-Based Multiple Access Schemes: Part 1—Air Interface Requirements,* Technical Report Tdoc 117/96, ETSI STC SMG2 Meeting No. 18, Helsinki, Finland, May 1996.

[15] van de Beek, J.-J., et al., *A Conceptual Study of OFDM-Based Multiple Access Schemes: Part 2—Channel Estimation in the Uplink,* Technical Report Tdoc 116/96, ETSI STC SMG2 Meeting No. 18, Helsinki, Finland, May 1996.

[16] van de Beek, J.-J., et al., *A Conceptual Study of OFDM-Based Multiple Access Schemes: Part 3—Performance Evaluation of a Coded System,* Technical Report Tdoc 166/96, ETSI STC SMG2 Meeting No. 19, Dusseldorf, Germany, September 1996.

[17] Linnartz, J. P. M. G., N. Yee, and G. Fettweis, "Multi-Carrier CDMA in Indoor Wireless Networks," *Conference Proceedings PIMRC'93,* Yokohama, Japan, September 1993, pp. 109–113.

[18] Prasad, R., *Universal Wireless Personal Communications,* Norwood, MA: Artech House, 1998.

[19] Prasad, R., and S. Hara, "Design and Performance of Multicarrier CDMA System in Frequency-Selective Rayleigh Fading Channels," *IEEE Trans. on Vehicular Technology,* Vol. 48, No. 5, September 1999, pp. 1584–1595.

[20] Prasad, R., and S. Hara, "An Overview of Multi-Carrier CDMA," *Proc. 4th IEEE Int. Symp. Spread Spectrum Techniques and Applications (ISSSTA'96),* September 1996, pp. 107–114.

[21] Prasad, R., S. Hara, and T. H. Lee, "BER Comparison of DS-CDMA and MC-CDMA for Frequency Selective Fading Channels," *Proc. 7th Tyrrhenian Intl. Workshop on Digital Communication,* September 1995, pp. 3–14.

[22] Prasad, R., and S. Hara, *Multiple Access Schemes Based on a Combination of Code Division and Multi-Carrier Techniques,* IEICE Technical Report, RCS96-48, June 1996.

[23] Takeda, D., et al., "Orthogonal Multicode OFDM-DS/CDMA System Using Partial Bandwidth Transmission," *IEICE Trans. Communication,* Vol. E81-B, No. 11, November 1998, pp. 2183–2190.

[24] Geraniotis, E. A., and A. Evaggelos, "Coherent Hybrid DS-SFH Spread Spectrum Multiple Access Communications," *IEEE JSAC SAC-3,* No. 5, September 1985, pp. 695–705.

[25] Proakis, J. G., *Digital Communications,* 3rd ed., New York: McGraw-Hill, 1995.

[26] Bello, P. A., "Characterization of Randomly Time-Variant Linear Channels," *IEEE Trans. on Communication Systems,* Vol. CS-11, 1963, pp. 360–393.

[27] Jakes, W. C., Jr., *Microwave Mobile Communications,* New York: John Wiley & Sons, 1974.

[28] Smulders, P. F. M., "Broadband Wireless LAN—A Feasibility Study," Ph.D. thesis, Eindhoven University of Technology, 1995.

[29] Witrisal, K., Y. H. Kim, and R. Prasad, "Frequency Domain Simulation and Analysis of the Frequency Selective Radio Channel for the Performance Analysis of OFDM," *Fachgesprach,* Braunschweig, Germany, September 1998.

[30] Tutucu, V., "Channel Estimation for OFDM System in Multipath Fading Environments for Wireless Broadband Communications," M.Eng. thesis, IRCTR-5-028-98, Delft University of Technology, Delft, the Netherlands, 1998.

[31] Zhao, Y., and A. Huang, "A Novel Channel Estimation Method for Mobile Communication Systems Based on Pilot Signals and Transform-Domain Processing," *IEEE Vehicular Technology Conference,* Vol. 3, 1997, pp. 2089–2093.

CHAPTER 8

A Practical OFDM System: Fixed Broadband Wireless Access (FBWA)

8.1 Introduction

At present wireline and cellular systems are suffering from a capacity and bandwidth crunch due to various economic, legal, and political causes [1, 2]. The delay in the deployment of 3G systems makes the situation even worse. Various technologies are being developed to fill the void of "last mile/hotspot" coverage based on fixed broadband wireless technology. These networks are typically configured in a cell-based point-to-multipoint topology. The deployment of these networks requires a high initial up-front investment to pay for base stations and backbone networks. Additionally, a significant number of potential subscribers may not be reached due to LOS constraints. This chapter proposes a solution based on OFDM wireless communication techniques to resolve the issue of coverage and capacity. The solution is based on proven technology and an offshoot of existing standards so that it is cost-effective to implement without much of an upfront cost in engineering and production. The proposal utilizes the existing IEEE 802.16 standard for WirelessMAN MAC and higher-layer interfaces to minimize non-PHY development costs [1–34].

The IEEE standards committee has recently standardized the IEEE 802.16 WirelessMAN air interface based on OFDM techniques to serve the needs of the BWA scheme primarily to build a metropolitan point-to-multipoint communications infrastructure. The air interface is designed to carry any type of data or multimedia traffic with full QoS support. The MAC supports burst FDD and TDD in a consistent framework. The application of OFDM technology in wireless applications has continued to grow and has become the forerunner for the next generation of cellular/wireless systems across the globe.

The requirements for the PHY of the proposed local coverage are different from those proposed in the IEEE 802.16 standard.

8.2 Motivation

It has been widely recognized that fixed broadband wireless systems are a viable alternative to overcoming the "last mile" distribution problem of bringing high-speed data connections to the end user. In areas where the cellular coverage and wired infrastructure are inadequate or nonexistent, fixed wireless solutions can be deployed much more quickly and at a fraction of the cost compared to wired networks.

There are currently several solutions being deployed or proposed to address this emerging fixed broadband wireless market. These solutions can be categorized based on various parameters including frequency band, range, bandwidth, throughput, regulatory and licensing requirements, and the like. Table 8.1 briefly summarizes the major fixed broadband wireless technologies.

The Local Multipoint Distribution Service (LMDS) has extremely high bandwidth; however, due to short range and LOS coverage limitations, it is not economically feasible for deployment for other than highly populated urban areas.

The Multichannel Multipoint Distribution System (MMDS), with its longer range, is a better match to cover medium-density areas, but it requires payment of a licensing fee.

The unlicensed spectrum allows for cost-effective immediate deployment and is attractive to unlicensed operators with limited financial resources.

8.2.1   Cell-Based Infrastructure

A large geographical area is covered using the cell-based infrastructure. One base station covers a predefined geographical area and handles the communications needs of the subscribers in that area of coverage. The nonoverlapping base stations are duplicated at regularly spaced intervals to provide wider coverage area with the frequency reuse principle. These base stations employ various medium-access mechanisms to coordinate and maintain the subscribers' communication links. This medium-access scheme can be based on TDMA, FDMA, CDMA or OFDM techniques. The base station provides controlled access to bandwidth allocation and sharing between subscribers.

8.2.1.1   Coverage Difficulties in Cell-Based Infrastructure

The disadvantage of the cellular system is that it requires a central location (like a mobile switching center for GSM/GPRS and UMTS, and so forth) through which all of the traffic must be channeled. This puts an unnecessary burden on the core and backbone networks because they need to be deployed for the maximum capacity envisioned. The system does not allow for scalability or provide an easy migration path for growth.

Table 8.1   Fixed Broadband Wireless Technology

Type	Frequency Band (GHz)	Bandwidth	Coverage (miles)
LMDS	28	1.2 GHz	3
MMDS	2.5	200 MHz	30
Unlicensed (ISM)	2.4	83 MHz	30
Unlicensed (UNII)	5.8	200 MHz	12

The cellular system also experiences coverage outages due to shadowing (loss of LOS) in high-rise office/residential and hilly areas. This has been overcome in certain circumstances by using smaller cells to provide coverage in fringe areas, but this adds costs for additional base stations and backbone networks.

8.2.2   Mesh Topology–Based Infrastructure

To provide the coverage in non-LOS (NLOS) areas, a "mesh" network topology has been employed. In this topology, each station is equipped with an omnidirectional antenna. Each station belonging to a network is capable of sending, receiving, and forwarding packets from any of its neighbors to various destinations using any of the possible routes. This topology does not require a backbone and can easily reach hidden locations through multiple hops.

There are some distinct disadvantages of this multipoint-to-multipoint network connectivity:

- The antenna is typically omnidirectional to cover the neighboring stations, which either reduces the range or decreases the power efficiency of the station.
- Each station has a good number of neighboring stations that can be reached with one hop. Although this is an advantage of mesh topology, which provides many alternate paths between any two points, the packet will reach not only the intended recipient, but all other unintended stations as well, possibly causing collisions.
- To avoid such collisions, there is a need for over-the-air coordination of packet routing of information, adding extra overhead to the packet. This distribution overhead reduces the achievable throughput of the system.

8.3   Proposed FBWA

The proposed solution is based on point-to-multipoint communications employing a mesh topology, which takes advantage of both cellular and mesh network topologies, but avoids the pitfalls of both. The proposed solution consists of a network topology with MAC protocol that facilitates the following:

- The network can be deployed incrementally as the need for coverage arises, thus providing scalability and an easy migration path to growth.
- The stations utilize the configurable directional (beam-forming) antenna, the width and direction of which can be configured through software control.
- Any new station can be added to the overall network infrastructure with easy coverage computation and link upgrade.
- Frequency and space diversity can be added to enhance the communication link and channel capacity if required.
- All the nodes in the network employ similar equipment.
- No expensive hubs or base stations are required.

• Backbone point-to-point links can be added later when a need for network expansion arises.

Figure 8.1 shows a simplified view of such a network topology. Each transceiver in the network infrastructure employs two types of interface, one for the subscriber's equipment interface (typical Ethernet) connectivity and the other for the RF antenna connectivity. This RF antenna has beam-forming capability, which can be configured under software control. For example the central site (node-1) antenna in the example could be configured as either omnidirectional or sectorized, depending on the infrastructure footprint, and the subscriber site antenna could be configured as directional, pointing toward the central site.

The initial installation may consist of as few as two transceivers: one at a central site (node-1) and the other at a subscriber site (any other nodes in Figure 8.1). In the initial installation, node-1 is equipped with either an omnidirectional or sectorized antenna, and the subscriber-side transceiver antenna is configured with a directional antenna pointing to the central site. When more subscribers need to be added with direct LOS to the central site, then they are installed with configuration of the antenna pointing to the central site as explained before.

When a new subscriber without direct LOS to the central station needs to be added (such as node-2 or node-3 in Figure 8.1), then any neighboring node in the vicinity of the subscriber site with direct LOS to the central site can be used as a relay station to connect to the central site. For example, node-2 could use either node-7 or

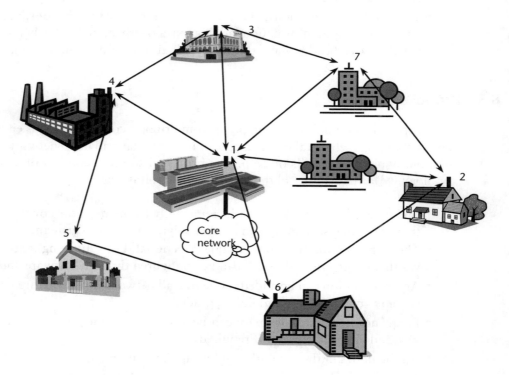

Figure 8.1   Single-hop point-to-multipoint network topology.

node-6 to reach the central site at node-1. For the purpose of explanation, it is defined that the central/serving site acts as the parent node and the served node as a child node. In this case node-6 or node-7 provides a hop for the NLOS connectivity to node-2. Node-2 is equipped with an antenna, which points either to node-6 or node-7, and the selected hop (repeater) node as a result needs to function both as a central site and a subscriber site; that is, it needs to time-share the connectivity between its child and parent nodes. Figure 8.2 shows the infrastructure antenna configuration for the central site and the parent and child nodes. For example node-7 in Figure 8.2 serves as both parent and child, so it needs to provide the functionality of both, and its antenna is multiplexed between omnidirectional mode for serving the child and directional mode for serving its own subscribers. As discussed later, the uplink connectivity is based on TDMA, but the downlink capability is based on broadcast TDM.

In case of expansion of the network, any node in the network can be promoted to become a hop node by reconfiguring the antenna port to switch between omnidirectional and directional modes, pointing to either a hop or a central site. This presents a very scalable and planned growth alternative, which reduces the upfront cost of network deployment and makes expansion very manageable.

The downlink flow of packets defines the parent-to-child node transmission, where as the uplink flow of packets is just the opposite. The uplink and downlink frequencies do not overlap and are assigned separate frequency bands. In this topology, there is no interference on downlink transmission because the child uses high-powered directional antenna for reception. Similarly, the uplink transmission from the child does not cause any interference because each child is power controlled to make it sure that adequate link quality is maintained and that each uses the directional antenna as well.

Frequency and antenna (directional) diversity allow simultaneous transmissions of many different frequency bands in the same geographical area. In any parent/

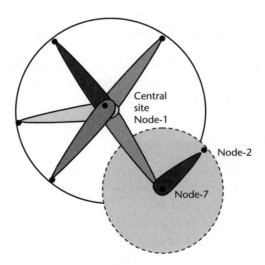

Figure 8.2 Network coverage and parent/child hop connectivity.

children group, the parent node services the children, using the adaptive TDMA scheme; thus, the available bandwidth is shared among the active child nodes. The MAC-layer performs these functions and synchronizes the hop station to multiplex between the role of child or parent at appropriate instants.

The packets can arrive at any node either via Ethernet interface or through the radio interface. Each transceiver maintains a map of the complete network infrastructure and routes the packet to the required destination in the most efficient manner.

8.4   Systems Requirements

This section specifies the requirements based on the proposed single-hop point-to-multipoint (SHPM) network topology for use in the FBWA application. The PHY shall provide digital, two-way voice, data, Internet, and multimedia services. It shall be compatible with traditional wireline, cable, DSL, and T1/E1 services.

To maximize the efficient utilization of spectrum in the unlicensed band, the air interface supports the multiplexing of subscribers using TDMA technology. The main features of the proposed PHY are as following:

- Duplex mode of operation (simultaneous Tx/Rx) using TDD;
- Full compatibility with IEEE 802.16 MAC;
- Uplink multiple access using TDMA;
- Downlink multiple access using broadcast TDM;
- Adaptive modulation and FEC coding in both up- and downlink;
- Support for smart antenna;
- High-speed asymmetric data-rate services;
- Dynamic allocation of bandwidth/channel;
- Allowance for handover and limited roaming;
- Easily adoption of cellular/wireline infrastructure.

This section primarily deals with the PHY requirements. The OFDM parameters are chosen based on these requirements. The systems PHY requirements are depicted in tabular form in Table 8.2.

Table 8.2   System PHY Parameters

Parameters	Value	Comments
Frequency bands	5.725–5.825 GHz	UNII band (USA) total BW = 100 MHz
Coverage range	≤10 miles	Last mile hotspot coverage
Delay spread tolerance	≤ 10 μs	Urban area
Channel BW	6 MHz	16 channels with end guard bands of 2 MHz
Data rate	2, 4, 8, 12 Mbps	

8.4.1 Parameter Selection

Based on the PHY requirements depicted in Table 8.2, OFDM-based systems parameter lists have been selected to fulfill the systems PHY requirements.

8.4.1.1 Selection of Guard Period

The OFDM-based fixed broadband wireless system needs to tolerate an RDS of about 10 μs. The guard period of 40 μs (four times the RDS) is chosen to mitigate any losses due to ISI.

8.4.1.2 Selection of OFDM Symbol Period

It is desirable to reduce to a minimum the throughput loss due to the addition of a guard period; therefore, a symbol period five to six times larger than the guard period is chosen. An OFDM symbol period of 240 μs (six times the guard period) is chosen, which results in less than 1-dB loss in throughput.

8.4.1.3 Selection of SC Spacing, Number of SCs, Modulation, and Coding Rate

Based on the guard period and OFDM symbol period, the spacing of SCs is computed so that they are orthogonal to each other. The SC spacing is computed as the inverse of the useful OFDM symbol period (1/(OFDM symbol period − guard period) = 1/(240 − 40) μs = 5 kHz).

The total channel bandwidth is allocated to 6 MHz. The parameter choice is tabularized in Table 8.3.

8.4.2 Communications Protocol

As illustrated in Figure 8.3, the PHY-layer protocol including upper-layer stack comprises two sublayers.

8.4.2.1 Physical Media–Dependent Sublayer

The physical media–dependent (PMD) sublayer involves the main processing parts of the PHY layer, including filtering equalization, synchronization, randomizing, FEC encoding/decoding and interleaving/deleaving, baseband pulse shaping, and other baseband processing units to enhance digitally modulated RF carriers over the air.

Table 8.3 BWAS Parameters List

Parameter	Number of SCs, 1,024; SCs Used, 960			
Data rate	2 Mbps	4 Mbps	8 Mbps	12 Mbps
Coding rate	1/2	1/2	3/4	2/3
Mapping/modulation	BPSK	QPSK	16-QAM	64-QAM
# Pilot insertion	16	16	16	16

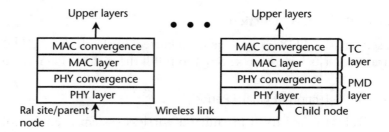

Figure 8.3   Proposed PHY layer with upper-layer protocol stack.

8.4.2.2   Transmission Convergence Sublayer

The transmission convergence (TC) sublayer is defined to adapt and map certain MAC services (such as changing resource allocations) to generic PMD services. These parts will be addressed at later stages of the development processes.

8.4.3   Duplex Schemes

In order to comply with IEEE 802.16.3 functional requirement [3], we propose to support both TDD and FDD systems and leave the selection of each system to the vendors or operators as they decide about implementation complexity, traffic scenario, cost objectives, and spectrum availability.

8.4.3.1   TDD

In TDD systems, the radio frame is divided into a downlink and an uplink section, offering flexible and dynamic allocation of uplink and downlink capacity. TDD enables the use of simpler antennas. In a BWA system, where the delay between transmission and reception can consist of a few time slots, a guard time between the downlink and uplink sections of the frames has to be introduced in order to avoid collision between time slots. However, the guard time reduces system throughput, especially if the system is designed for low latency.

8.4.4   Downlink Transmission

8.4.4.1   Downlink Multiple-Access Scheme

Each downlink RF channel (e.g., 6 MHz wide) is subdivided into fixed frames with which the RF carrier is suitability modulated (e.g., BPSK, QPSK, 16-QAM, 64-QAM) to provide a digital bit stream (e.g., 2 to 12 Mbps). Within each RF channel a frame structure is used to organize and schedule the transmission of voice, video, and data traffic.

8.4.4.2   Modulation Scheme

The applicable modulation schemes for the downlink are BPSK, QPSK, 16-QAM, or 64-QAM. Adaptive modulation and coding shall be supported in the downlink. The uplink shall support different modulation schemes for each user, based on the MAC burst configuration messages coming from the central site.

8.4.4.3   Downlink Randomization, Channel Coding and Interleaving, Symbol Mapping, and Baseband Shaping

The downlink channel supports various modulation formats and FEC coding on the user data portion of the frame. Different modulation formats and FEC groups can be defined on a subscriber-level basis. In this way the downlink channel supports adaptive modulation and coding. Note that each frame contains a control portion with fixed modulation (QPSK) and an FEC scheme.

8.4.4.4   Randomization for Spectrum Shaping

Prior to FEC encoding, the downlink channel is randomized to ensure sufficient bit transitions to support clock recovery and to minimize the occurrence of unmodulated carrier frequencies. This process is done by modulo-2 addition (XOR) of the data with the output of the linear-feedback shift register (LFSR) with the characteristic polynomial $1 + X^{14} + X^{15}$. The LFSR is cleared and preset at the beginning of each burst to a known value. The preambles are not randomized, and only information bits are randomized. The LFSR sequence generator pauses while parity bits are transmitted.

8.4.4.5   Downlink Channel FEC Definitions

The structure of the 802.16 MAC supports several FEC coding schemes (such as block turbo coding and concatenated Reed-Solomon and convolutional coding), but for the purpose of this application, only convolutional code is employed. In addition, the provision of suppressing all FEC and operation using the ARQ mechanism in the 802.16 MAC for error control is also included.

Convolutional Code
The convolutional code can be configured to code rates of 1/2, 2/3, and 3/4 using a puncturing convolutional code constraint length of $K = 12$. The interleaving is supported across the duration of one symbol.

8.4.4.6   Symbol Mapping

The mapping of bits into in-phase and quadrature-phase is Gray coded. The mapping of bits into symbols is based on the data rate and modulation scheme used.

8.4.4.7   Baseband Pulse Shaping

Prior to up conversion, in-phase and quadrature-phase signals are spectrally shaped to conform to the channel bandwidth requirement using a square-root raised cosine filter. The roll-off factor shall be either 0.15 or 0.25. The ideal square-root cosine is defined by the following TF:

$$H(f) = 1 \quad for |f| < f_s(1-\alpha)$$

$$H(f) = \left\{ \frac{1}{2} + \frac{1}{2}\sin\left(\frac{\pi(f_s - |f|)}{2\alpha f_s}\right) \right\}^{0.5} \quad for\ f_s(1-\alpha) \le f_s(1-\alpha)$$

$$H(f) = 0\ for\ |f| \ge f_s(1-\alpha)$$

where

$f_s = \dfrac{1}{T_s} = \dfrac{R_S}{2}$ is the Nyquist sampling frequency;

Ts is the modulation symbol duration.

8.4.4.8   Frequency Domain Equalization Scheme

The BWAS systems employing 2–11-GHz spectrum may operate in NLOS conditions in which severe multipath is encountered. Multipath delay spread is a major transmission problem, which affects the design of modulation and equalization. Delay spread varies with the environment and characteristics of transmit and receive antennas. In typical MMDS operating conditions, average delay spread is ~0.5 μs; only 2% of measured delay spreads are greater than ~8–10 μs [4–6].

8.4.5   Uplink Transmission

8.4.5.1   Uplink Multiple Access

The uplink multiple-access method shall be TDMA.

8.4.5.2   Uplink Modulation Format

The uplink modulation shall be BPSK, QPSK, 16-QAM, or 64-QAM.

8.4.5.3   Uplink Randomization, Channel Coding and Interleaving, Symbol Mapping, and Baseband Shaping

The uplink channel has processing units similar to those described for the downlink. However, greater flexibility in packet transmission is allowed. The subscriber stations are transmitting only after receiving some configuration information from the central site through MAC messages. Several different configurations can be adjusted on the uplink channel on a burst-to-burst basis. The uplink payload is segmented into blocks of data designed to fit into the proper code word size (including TC sublayer, TC header). Note that payload length may vary from burst to burst.

Randomization for Spectrum Shaping
The uplink modulator uses a randomizer using LFSR with the characteristic polynomial $1 + X^{14} + X^{15}$, with a 15-bit programmable seed. At the beginning of each burst, the register is cleared and the seed value is loaded. The seed value is used to calculate the scrambler output bit, obtained as the XOR of the seed with first bit of data of

each burst [which is the most significant bit (MSB) of the first symbol following the last symbol of the preamble].

FEC Schemes for the Uplink Channel
The convolutional code can be configured to code rates of 1/2, 2/3, and 3/4 using a puncturing convolutional code constraint length of $K = 12$.

Interleaving for the Uplink Channel
Interleaving is applied for the uplink channel only with a convolutional FEC scheme. The interleaving is supported across the duration of one symbol.

Baseband Pulse Shaping
Using a square-route raised cosine filter, in-phase and quadrature-phase signals are separately shaped to conform to the channel bandwidth requirement prior to up-conversion. Either 0.15 or 0.25 will be the roll-off factor.

8.4.6 Frame Structure

The frame format is simple and block-based and accommodates TDD operation. Figure 8.4 illustrates downlink TDD operation. Figure 8.5 illustrates uplink operation. Several features characterized the frame formatting. Key among these is the use of preamble sequence (PS) symbols, which occur at an interval of every N symbols (on the downlink). The PS acts as a training sequence and is known to both the transmitter and receiver. The training sequence can be used for frequency/carrier phase tracking, time tracking, and channel equalization on the downlink. The periodic nature of the training sequence allows for initial acquisition and tracking during symbol reception. The downlink transmission is continuous; it carries the payload for several subscribers and broadcasts a dummy load with the training

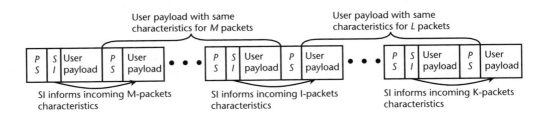

Figure 8.4   Downlink TDD frame structure.

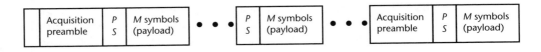

Figure 8.5   Frame structure for uplink.

sequence when there are no subscribers, which allows the idle subscriber to obtain system information and to continue to track the serving site.

The training sequence is selected to have good correlation properties (i.e., its autocorrelation resembles an IR), so that the transmission channel response can be identified with high accuracy. For easy demodulation even in low SNR cases (and optimal correlation sequence choices), the sequence is derived from a QPSK alphabet.

As Figures 8.4 and 8.5 illustrate, the PSs sandwich the data of length N on the downlink or length M on the uplink. This property is particularly useful for FD equalization because these PS segments on each end of the N (or M) point data serve as cyclic prefixes, so that the DFT in the FD behaves periodically. The DFT would be taken over the duration of the PS on both ends of an N (or M) user payload. Note that this usage is very similar to the cyclic prefix used in OFDM. However, here the PS is used as both a pilot symbol, as well as the cyclic prefix, and this results in a higher spectral efficiency compared to other OFDM systems.

As mentioned earlier, this PS is used for estimating the frequency, time, and channel IR during acquisition and tracking. A particular advantage over OFDM-type approaches is that these pilot symbols span the whole frequency spectrum and, thus, are less affected by the notches due to multipath (frequency selectivity) of the channel. A DFE or other TD equalizer can also exploit these known pilot sequences to improve both their channel estimation and equalization performance. In particular, the use of periodic known sequences can reduce the propagation of decision errors within a DFE.

Another feature of the proposed framing is the system information (SI) field of length S, carried at required intervals for the indication of the modulation type, the coding rate of the user payload, and subscriber-specific information (such as whether the next payload is intended for them or not), which follows the next PS. Note that the interval for a given modulation may span several packets, and the minimum duration of a given modulation type, that is, the spacing between S fields, is a network-specific parameter.

For the TDD option, a dummy payload packet called Idle Packet is used to inform the subscribers that following packet segments do not carry any subscriber-specific payload. The uplink capacity is shared among the active subscribers, who have requested the services from the central site. In the TDD mode, the central site continues to broadcast the PS followed by a dummy packet during the idle period, even though no uplink bandwidth grants are made. The subscribers can use the idle packets to synchronize with the central site and may request the desired services.

8.4.7 MAC Consideration and Conformance

As mentioned earlier, for cost-effective and quick deployment reasons, requirements for the fixed broadband wireless application shall strive to fit into the 802 system model [1, 7]. The 802.16 MAC supports universal 48-bit addresses and MAC multicast in the downlink direction only, not uplink. The 802.16 protocols support 802.1 bridging services and protocols, including support of the virtual LAN tag and priority ID [8–10]. The 802.16 protocols support encapsulation of 802.2 [logical link control (LLC)] by the MAC protocol.

8.4.7.1 MAC Functional Requirements

The following describes the functional requirements to be performed by the wireless MAC. In conjunction with the PHY equipment, the MAC assures that QoS requirements for the wireless segment are met, such as delay, delay variation, and the like. This does not include the detailed list of the MAC functionalities, but some important control and supervisory requirements.

Link Acquisition
The following are tasks for link acquisition:

- Download to subscriber the local channel plan, data-rate options, modulation options, FEC types, and time-slot arrangement employed at a specific cell;
- Establish link at proper uplink power and frequency;
- Provide time-slot timing calibration.

Link Maintenance
Link maintenance requirements are the following:

- Provide uplink power control and frequency control (optional) to maintain specified error-rate performance during link dynamics such as rain fades;
- Provide time-slot timing control;
- Provide interference detection and mitigation;
- Provide redundant hardware control.

Resource Allocation
Some important control and supervisory requirements are to:

- Provide admission control for connections based on available resources;
- Provide dynamic allocation of channels and time slots according to traffic and traffic-priority requirements;
- Police traffic conflicts;
- Provide buffer management.

Link Monitoring
The following describes the functional requirements to be performed by link monitoring:

- Provide status of link performance (error seconds);
- Provide status of hardware;
- Maintain status of bandwidth and resource availability;

• Provide fault detection, isolation, and correlation.

8.4.8 Adaptability of the Proposed Kernel Architecture to BWAS

The baseband transceiver architecture for the BWAS application is very similar to the WLAN 802.1 la transceiver design requirements, except for the length of the FFT/IFFT block, symbol, and guard period. Since the FFT/IFFT kernel allows for the configurability of the FFT/IFFT length, the WLAN 802.11a development platform could be utilized for the BWAS application, as well for minor modifications in timing block. The MAC requirements for BWAS are different from the requirements of the IEEE 802.11 [1] protocol, but as was emphasized earlier in the design assumptions and ground rules, the BWAS application requirements for MAC shall be compatible to the IEEE 802.16 MAC specifications [7]. The main goal of this chapter has been to show the flexibility of the transceiver design based on the object-oriented approach with the wireless-specific design of kernel.

8.4.9 Summary

The fixed BWAS based on the OFDM scheme was proposed and requirements were set to complement the wireline and wireless infrastructure in last-mile and hotspot coverage to provide a low-cost, quickly deployable solution. The proposed systems were designed; the PHY parameters and the access techniques for uplink and downlink transmission were formulated based on the OFDM scheme. The concept of a point-to-multipoint mesh network topology was proposed to mitigate the NLOS situation, and the requirements for central, parent, and child site transceiver antennas were set. The choice of beam-forming antenna under software control allowed the promotion of any node to a parent node and enabled nodes to serve as hop or relay agents for neighboring nodes without direct LOS to the central station. This scheme eliminated the need for multiple transceiver units at a node, which need to be upgraded as hop or relay nodes, thus facilitating the deployment of homogeneous components in network buildup. This scheme also provided an easy path for migration, and no derivative of the transceiver component was required. Finally, it was shown that the design based on the proposed architecture using the distributed kernels utilizing the object-oriented programming principles allowed the quick development of the BWAS transceiver.

8.5 Ubiquitous Connectivity

The quest for communication anywhere, any time has fuelled an enormous growth in the realm of cellular/wireless communications around the world. Wireless network applications have been compartmentalized into three basic categories, primarily based on coverage and mobility requirements around the world.

However, these standards were not built to exploit commonality or to promote coexistence and were formulated in isolation. For truly ubiquitous connectivity, a mobile user needs to carry several wireless devices, depending on whether he or she is at home, in the office, or on road, as shown in Figure 8.6. Additionally, the user has to look continuously for available wireless services in the immediate vicinity and manually switch to them. This causes abandonment of the present communication thread, which results in wasted time and resources and causes extreme inconvenience and overhead. However, looking at the scenario, it seems obvious that there is a need for a wireless device that can provide ubiquitous connectivity irrespective of a user's location, mobility, and throughput needs and that can provide seamless roaming across standards without user intervention.

The need for instant connectivity irrespective of locality and mobility is very desirable; otherwise, one has to cope with the situation discussed above. The architecture based on distributed kernels and localization of waveform-specific signal processing with emphasis on object-oriented programming can be extended to other wireless systems as well. The mobile device designed with this architecture can support more than one wireless protocol with little or no additional resources or overhead. The development platform based on this architecture can be dubbed the "universal wireless platform" because it allows for easy migration and adaptation to various wireless applications. This also simplifies the logistics and maintenance issues of product development. The manufacturers of mobile appliances need not worry about maintaining the plethora of various product derivatives, engineering changes, and inventory lists. This needs to be proven and tried for multiprotocol wireless devices in the field. Here are the recommendations to enrich the offering of this architecture:

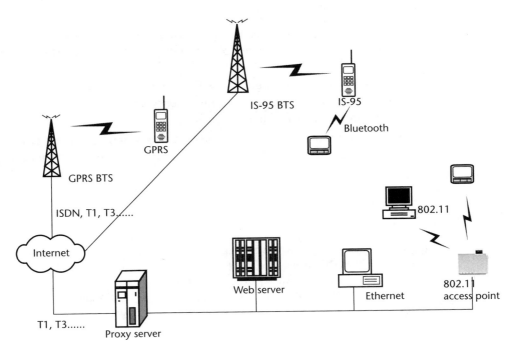

Figure 8.6 Various modes of wireless connectivity in Internet and intranet environments.

- Extension of this architecture and development methodology to other wireless protocol applications;
- Development of a standard user interface for interoperability, interchangeability, and compatibility;
- Standardization of a new class of processing device that is wireless-specific and signal-processing-centric.

The role of technology in society is to lessen the burden of daily drudgery in human life. The acceptance of a technology depends on how it helps the common people in managing their daily chores. Wireless-based connectivity will remain fragmented and very specific based on locality, mobility, and business issues. This means that there is a need for a multiprotocol mobile appliance that can unite this fragmented wireless connectivity and provide seamless roaming to the wireless world. Seamless wireless connectivity requires a wireless platform capable of accommodating many wireless protocols at low cost and with minimal resource overhead and capable of switching between them as desired. Additionally, it needs the support of the higher-layer protocols to assist in switching and other systems-related decisions making. The issue of the control and maintenance of seamless roaming needs to be resolved separately since it involves the coordination from the higher layers (such as RLC, LLC, and MAC) of the protocol. This coordination may involve the PHY to assist in searching for alternate wireless services in the vicinity, establishing link quality, dealing with congestion, billing, and other related issues. The following future enhancements for continued acceptance and growth of this architecture are very desirable:

- Design of a universal wireless platform, which can accommodate current slews of wireless protocols and be flexible enough to accommodate the evolving protocols;
- Harmonizing the upper-layer (above PHY) protocols across the access scheme and protocols to ease resource requirements and delay in information flow from one protocol to another;
- Standardization and harmonization of various wireless protocol layers to ensure compatibility, interchangeability, and interoperability across various platforms and interfaces.

References

[1]   Jha, U., "Wireless Landscape—A Need for Seamless Connectivity," *Wireless Personal Communication (WIRE) Journal,* Vol. 22, No. 2, August 2002, pp. 275–283.

[2]   Jha, U. S., et al., "Novel Application-Specific Signal Processing Architectures for Wideband CDMA and TDMA Applications," *VTC 2000*, Tokyo, Japan, 2000.

[3]   IEEE802.16.3-00/02r4, Functional Requirements for the 802.16.3 Interoperability Standard, dated 2000-09-22.

[4]   Porter J. W., and J. A. Thweatt, "Microwave Propagation Characteristics in the MMDS Frequency Band," *Proc. Intl. Conf. on Communications,* New Orleans, LA, June 2000, pp. 1578–1582.

[5]   Sari, H., G. Karam, and I. Jeanclaude, "Transmission Techniques for Digital Terrestrial TV Broadcasting," *IEEE Communications Magazine,* Vol. 33, No. 2, February 1995, 100–109.

[6]   Tarokh, V., and H. Jafarkhani, "On the Computation and Reduction of the Peak-to-Average Ratio in Multicarrier Communications," *IEEE Trans. on Communications,* Vol. 48, No. 1, January 2000, pp. 37–44.

[7]   IEEE 802.16ab-01/01rl, Air Interface for Fixed Broadband Wireless Access Systems, Part A: Systems between 2–11 GHz, July 2001.

[8]   ISO 15802-3:1998, Information Technology—Telecommunications and Information Exchange between Systems—Local and Metropolitan Area Networks—Common Specifications—Part 3: Media Access Control (MAC) Bridging.

[9]   ISO/IEC 15802-5:1998, Information Technology—Telecommunications and Information Exchange between Systems—Local and Metropolitan Area Networks—Common Specifications—Part 5: Remote Media Access Control (MAC) Bridging.

[10] IEEE 802. IF-1993, IEEE Standards for Local and Metropolitan Area Networks: Common Definitions and Procedures for IEEE 802 Management Information.

[11] LAN/MAN Standards Committee of the IEEE Computer Society, Part 11: Wireless LAN Medium Access Control (MAC) and Physical Layer (PHY) Specifications—High Speed Physical Layer in the 5 GHz Band, ANSI/IEEE Std 802.11, 1999 Edition.

[12] Jha, U., "Low Complexity Resource Efficient OFDM Transceiver Design," *ICPWC-2002,* New Delhi, India, December 15–18, 2002.

[13] Jha, U. S., "Wireless Specific DSP Implementation, Wireless System Design Issues," *Wescon '98,* Anaheim, CA, September 15–17, 1998, pp. 234–241.

[14] Bolle, A., O, Eriksson, and A. Nascimbene, "Competitive Broadband Access Via Microwave Technology," *Ericsson Review,* No. 4, 1998, pp. 161–171.

[15] van Nee R., and R. Prasad, *OFDM for Wireless Multimedia Communications,* Norwood, MA: Artech House, 2000.

[16] Falconer, D., and S. L. Ariyavisitakul, "Modulation and Equalization Criteria for 2–11 GHz Fixed Broadband Wireless Systems, IEEE802.16c-00/13," *IEEE 802.16 Meeting,* Denver, CO, September 11–15, 2000.

[17] Erceg, V., et al., "A Model for the Multipath Delay Profile of Fixed Wireless Channels," *IEEE JSAC,* Vol. 17, No. 3, March 1999, pp. 399–410.

[18] Erceg, V., et al., "An Empirically Based Path Loss Model for the Wireless Channels in Suburban Environments," *IEEE JSAC,* Vol. 17, No. 7, July 1999, pp. 1205–1211.

[19] Hari, K. V. S., and K. P. Sheikh, "Interim Channel Model for G2 MMDS Fixed Wireless Applications," IEEE 802.16.3c-00/4992, *IEEE Plenary Meeting,* Tampa, FL, November 6–9, 2000.

[20] Forney, D., "Burst Correcting Codes for the Classic Bursty Channel," *IEEE Trans. Communications,* Vol. COM-19, No. 5, October 1971, pp. 772–781.

[21] ETSIEN301210V1.1.1 1999.

[22] Hewitt, E., "Turbo Product Codes for LMDS," *IEEE Radio and Wireless Conference,* August 1998.

[23] Hagenauer, J., and P. Hocher, "A Viterbi Algorithm with Soft-Decision Outputs and Its Applications," *IEEE Globecom 1989,* November 1989, pp. 1680–1685.

[24] Drury, G., G. Markarian, and K. Pickavance, *Coding and Modulation for Digital Television,* Boston, MA: Kluwer Academic Publishers, 2001.

[25] Hagenhauer, J., and E. Offer, "Iterative Decoding of Binary Block and Convolutional Codes," *IEEE Trans. Inf. Theory,* Vol. IT-42, No. 2, March 1996, pp. 429–445.

[26] Van den Bos, C., M. H. L. Kouwenhoven, and W. A. Serdijn, "The Influence of Nonlinear Distortion on OFDM Bit Error Rate," *Proc. Intl. Conf. on Communication,* New Orleans, LA, June 18–22, 2000, pp. 1125–1129.

[27] Dark, M. V., "Adaptive Frequency-Domain Equalization and Diversity Combining for Broadband Wireless Communications," *IEEE JSAC,* Vol. 16, No. 8, October 1998, pp. 1385–1395.

[28] Falconer, D., and S. L. Ariyavisitakul, "Frequency Domain Equalization for 2–11 GHz Fixed Broadband Wireless Systems," *Tutorial, Session 11 of IEEE 802.16,* Ottawa, Canada, January 22, 2001.

[29] Aue, V., G. P. Fettweis, and R. Valenzuela, "A Comparison of the Performance of Linearly Equalized Single Carrier and Coded OFDM over Frequency Selective Fading Channels Using the Random Coding Technique," *Proc. ICC'98,* June 7–11, 1998, pp. 753–757.

[30] Czylwik, A., "Comparison between Adaptive OFDM and Single Carrier Modulation with Frequency Domain Equalization," *Proc. VTC'97,* Phoenix, AZ, May 1997, pp. 865–869.

[31] Kadel, G., "Diversity and Equalization in Frequency Domain—A Robust and Flexible Receiver Technology for Broadband Mobile Communications Systems," *Proc. VTC'97,* Phoenix, AZ, May 4–7, 1997, pp. 894–898.

[32] Gusmao, A., et al., "Comparison of Two Modulation Choices for Broadband Wireless Communications," *Proc. VTC'00,* Tokyo, Japan, June 2000, pp. 1300–1305.

[33] EN301-210, Digital Video Broadcasting (DVB); Framing Structure, Channel Coding and Modulation for Digital Satellite News Gathering (DSNG) and Other Contribution Applications by Satellite, ETSI, 1998-12.

[34] IEEE 802.1Q-1998, IEEE Standards for Local and Metropolitan Area Networks: Virtual Bridged LANs.

About the Author

Ramjee Prasad received his B.Sc. in engineering from the Bihar Institute of Technology, Sindri, India, and his M.Sc. in engineering and Ph.D. from Birla Institute of Technology (BIT), Ranchi, India, in 1968, 1970, and 1979, respectively.

Professor Prasad joined BIT as a senior research fellow in 1970 and became an associate professor in 1980. While he was with BIT, he supervised a number of research projects in the area of microwave and plasma engineering. From 1983 to 1988, he was with the University of Dar es Salaam (UDSM), Tanzania, where he became a professor of telecommunications in the Department of Electrical Engineering in 1986. At UDSM, he was responsible for the collaborative project "Satellite Communications for Rural Zones" with Eindhoven University of Technology, the Netherlands. From 1988 through 1999, he was with the Telecommunications and Traffic Control Systems Group at Delft University of Technology (DUT), where he was actively involved in the area of wireless personal and multimedia communications (WPMC). He was the founding head and program director of the Center for Wireless and Personal Communications (CWPC) of the International Research Center for Telecommunications—Transmission and Radar (IRCTR). Since 1999, Professor Prasad has been with Aalborg University. He has acted as the codirector of the Center for PersonKommunikation (CPK) and the research director of the Department of Communications Technology, and is the founding director of the Center for Teleinfrastructure (CTIF). Professor Prasad holds the chair of wireless information and multimedia communications. He was involved in the European ACTS project "Future Radio Wideband Multiple-Access Systems" (FRAMES) as a DUT project leader. He is a project leader of several international, industrially funded projects. He is the project coordinator of the European sixth framework integrated project "My Personal Adaptive Global NET" (MAGNET). Professor Prasad has published over 500 technical papers, contributed to several books, and has authored, coauthored, and edited 14 books: *CDMA for Wireless Personal Communications, Universal Wireless Personal Communications, Wideband CDMA for Third Generation Mobile Communications, OFDM for Wireless Multimedia Communications, Third Generation Mobile Communication Systems, WCDMA: Towards IP Mobility and Mobile Internet, Towards a Global 3G System: Advanced Mobile Communications in Europe, Volumes 1 and 2, IP/ATM Mobile Satellite Networks, Simulation and Software Radio for Mobile Communications, Wireless IP and Building the Mobile Internet, WLANs and WPANs towards 4G Wireless, Technology Trends in Wireless Communications*, and *Multicarrier Techniques for 4G Mobile Communications*, all published by Artech House. His current research

interests lie in wireless networks, packet communications, multiple-access protocols, advanced radio techniques, and multimedia communications.

Professor Prasad has served as a member of the advisory and program committees of several IEEE international conferences. He has also presented keynote
speeches and delivered papers and tutorials on WPMC at various universities, technical institutions, and IEEE conferences. He also participated in the European cooperation in the scientific and technical research (COST-231) project dealing with the
evolution of land mobile radio (including personal) communications as an expert for
the Netherlands, and he was a member of the COST-259 project. Professor Prasad
was the founder and chairman of the IEEE Vehicular Technology/Communications
Society Joint Chapter, Benelux Section, and is now the honorary chairman. In addition, Professor Prasad is the founder of the IEEE Symposium on Communications
and Vehicular Technology (SCVT) in Benelux, and he was the symposium chairman
of SCVT'93.

In addition, Professor Prasad is the coordinating editor and editor-in-chief of the
Kluwer International Journal on Wireless Personal Communications and a member
of the editorial board of other international journals, including the *IEEE Communications Magazine* and *IEE Electronics Communication Engineering Journal*.
He was the technical program chairman of the PIMRC'94 international symposium held in The Hague, the Netherlands, from September 19–23, 1994, and
also of the Third Communication Theory Mini-Conference in Conjunction with
GLOBECOM'94, held in San Francisco, California, from November 27–30, 1994;
the conference chairman of the Fiftieth IEEE Vehicular Technology Conference and
the steering committee chairman of the Second International Symposium WPMC,
both held in Amsterdam, the Netherlands, from September 19–23, 1999; and the
general chairman of WPMC'01, which was held in Aalborg, Denmark, from September 9–12, 2001. He is also the general chairman of the First International Wireless Summit (IWS 2005) to be held in Aalborg, Denmark, on Septmber 17–22, 2005.

Professor Prasad is also the founding chairman of the European Center of
Excellence in Telecommunications, known as HERMES. He is a fellow of IEE, a
fellow of IETE, a senior member of IEEE, a member of the Netherlands Electronics
and Radio Society (NERG), and a member of IDA (Engineering Society in
Denmark).

Index

The Artech House Universal Personal Communications Series

Ramjee Prasad, Series Editor

For further information on these and other Artech House titles,
including previously considered out-of-print books now available through our
In-Print-Forever® (IPF®) program, contact:

Artech House
685 Canton Street
Norwood, MA 02062
Phone: 781-769-9750
Fax: 781-769-6334
e-mail: artech@artechhouse.com

Artech House
46 Gillingham Street
London SW1V 1AH UK
Phone: +44 (0)20 7596-8750
Fax: +44 (0)20 7630-0166
e-mail: artech-uk@artechhouse.com

Find us on the World Wide Web at:
www.artechhouse.com